NEW DEVELOPMENTS IN PROCESSING OF SLUDGES AND SLURRIES

Proceedings of A Round-Table seminar organized by the Commission of the European Communities, Directorate-General Science, Research and Development, Environment Research Programme, held in Bilthoven, The Netherlands, 22 April 1985

NEW DEVELOPMENTS IN PROCESSING OF SLUDGES AND SLURRIES

Edited by

A. M. BRUCE

Water Research Centre, Stevenage, UK

P. L'HERMITE

Commission of the European Communities, Brussels, Belgium

and

P. J. NEWMAN

Water Research Centre, Medmenham, UK

ELSEVIER APPLIED SCIENCE PUBLISHERS
LONDON and NEW YORK

ELSEVIER APPLIED SCIENCE PUBLISHERS LTD
Crown House, Linton Road, Barking, Essex IG11 8JU, England

Sole Distributor in the USA and Canada
ELSEVIER SCIENCE PUBLISHING CO., INC.
52 Vanderbilt Avenue, New York, NY 10017, USA

WITH 22 TABLES AND 29 ILLUSTRATIONS

© ECSC, EEC, EAEC, BRUSSELS AND LUXEMBOURG, 1986

British Library Cataloguing in Publication Data

New developments in processing of sludges
and slurries.
1. Agricultural wastes   2. Slurry
I. Bruce, A. M.   II. L'Hermite, P.
III. Newman, P. J.
628'.746   TD930

ISBN 1-85166-011-9

Library of Congress CIP data applied for

Publication arrangements by Commission of the European Communities, Directorate-General Information Market and Innovation, Luxembourg

EUR 10359

LEGAL NOTICE
Neither the Commission of the European Communities nor any person acting on behalf of the Commission is responsible for the use which might be made of the following information.

Printed in Great Britain by Galliard (Printers) Ltd, Great Yarmouth

PREFACE

This book contains the proceedings of a 'round-table' meeting held at Bilthoven, The Netherlands, in April 1985 under the auspices of the Commission of the European Communities' Concerted Action (COST 681) on 'the treatment and use of organic sludges and liquid agricultural wastes'. Working Party No. 1 of the Concerted Action has the task of coordinating nationally funded research on the processing of sewage sludges and animal slurries. It also serves to promote regular exchanges of information on relevant research among the 15 countries participating in the COST 681 activity. The object of this 'round-table' was to review and discuss new research developments in the field of sludge and slurry processing. Some of the research described is still at a fairly early stage, but in other cases it has reached the phase of practical application.

Three of the papers relate to sludge dewatering methods, two relate to methods of aerobic stabilization of sludge and another paper describes a novel development in the conversion of sewage sludge to a usable fuel. The results of examining the influence of sludge processing techniques on some of the important organic micropollutants in sewage sludges are described in other contributions. The remaining paper deals with research on new ways of processing animal manures. The wide variety of research topics covered are critically reviewed and assessed by two members of Working Party No. 1—Professor Tom Casey and Professor Peter Balmer.

It is expected that the information presented will be of value both to other researchers in the field of sludge and slurry processing and, equally important, to the practitioners who are involved with the design and operation of processing plant.

A. M. BRUCE

vii

CONTENTS

Conclusions

<u>SESSION I</u>

CHAIRMAN: Mr. A.M. BRUCE

Rapporteur: Dr. P.J. NEWMAN

APPLICATION OF ELECTRICAL FIELDS TO THICKEN AND DEWATER SEWAGE SLUDGES

F. COLIN
Scientific Manager
Institut de Recherches Hydrologiques (I.R.H.)
NANCY - France

Summary

Benefits of applying electric fields for solid/liquid separation in urban sewage sludges. Electro-chemistry of interfaces and definition of electro-kinetic effects. Basic laws of electrophoresis and of electro-osmosis.
Application of electric fields:
- to speed up sedimentation and gravity thickening of sludges;
- to concentrate sludges by filtering without any cake production;
- to dewater liquid sludges until a solid residue is produced.
For each of the mechanisms of the design of the experimental devices and of projects, known performance data and power consumption.
Discussion of the benefits of such techniques and identification of the most promising field of application.

1. INTRODUCTION

Sewage sludges are made up of small solid particles suspended in an aqueous solution. Within this, the solid phase is extremely finely dispersed and therefore presents an extremely large developed surface in contact with the liquid. Under these conditions, surface phenomena play a decisive role in determining the behaviour of the systems and, in particular, the prospects of separating the solid and liquid phases.

The surface of a solid immersed in an aqueous solution acquires an electrical charge. An electrical double layer, consisting of a part fixed to the solid plus a mobile diffuse part, forms at the interface between the two phases. Consequently, application of an electric field acting selectively on the mobile and fixed parts of the electrical ionic double layer induces the solid and liquid phases to move relative to each other. Conversely, forced displacement of the solid in relation to the liquid or vice versa in turn generates an electric field at the centre. All phenomena of this type are known as "electro-kinetic effects".

This paper demonstrates the prospects for utilising these phenomena in practice to enhance thickening or dewatering of sewage sludges and outlines some of the methods which might come into consideration. Although the fundamental phenomena have long been known, there have been very few full-scale applications. This paper attempts to explain the reasons why.

2. ELECTRO-CHEMISTRY OF THE INTERFACES AND ELECTRO-KINETIC EFFECTS

2.1 Origin of the electrical charge at the interface

A variety of physical and chemical mechanisms can induce electrical charges at the surface of solids immersed in aqueous solution:

(i) Ionic dissociation of groups carried to the surface of the solid by the molecules. In general, this form of dissociation is in equilibrium, with the degree of dissociation depending on the pH. Proteins are one example:

$$HOOC-[protein]-NH_3^+ \rightleftharpoons HOOC-[protein]-NH_2 \rightleftharpoons {}^-OOC-[protein]-NH_2$$

pH < isoelectric point   pH = isoelectric point   pH > isoelectric point
solid bears a positive charge   solid bears no electric charge   solid bears negative charge

The pH at which there is no dissociation and the surface of the solid bears no electrical charge is known as the "isoelectric point".

(ii) Replacement of ions in the crystal lattices by ions of different valency thus producing a surplus or deficit in the electrical charge of the solid;

(iii) preferential adsorption of selected ions (usually anions) at the surface of the solid.

2.2 Composition of the electrical ionic double layer

The electrical charges attached to the surface of the solid interact with the anions and cations in the solution, attracting one category and repelling the other, to form an electrical double layer consisting of two parts of opposite sign (equivalent to the two plates of a capacitor). This arrangement can be disturbed by agitation by heat or with increasing distance from the surface of the solid.

Several authors have studied the fundamental electro-chemical properties of the electrical double layer, including Helmholtz (1), Stern (2), Gouy (3), Chapman (4) and Grahame (5), each of whom have produced increasingly complex, sophisticated models.

The inner part of the double layer is strongly attached to the solid, whereas the outer part is mobile and far more diffuse. The two are separated by a boundary plane at an unknown position. The ions in the double layer are hydrated and, therefore, carry water molecules which may be regarded either as attached to the solid or as mobile in the solution, depending on the case in point.

In practice, therefore, a distinction must be drawn between suspended particles bearing, on their surface, ions solvated by water molecules firmly attached to the solid on the one hand and the bulk of the solution containing mobile solvated ions on the other. These two phases are separated by the boundary plane which, in effect, is the outer sheath around the particles and the matter to which they are attached. Consequently, it is easy to understand why the properties of the matter at this interface determine the behaviour of the suspended particulates. The most important of all the properties of this plane is its electrical potential, compared with the potential of the liquid. This is known as its zeta potential. The zeta potential determines the forces of the electro-kinetic effects.

2.3 Definition of the electro-kinetic effects

In essence, there are four electro-kinetic effects, depending whether (i) the solid is mobile, in relation to the liquid, or (ii) vice versa or whether (iii) an electric field is imposed to force the solid to move in relation to the liquid or (iv) on the contrary, the movement comes first and induces an electric field.

Electric field \ Mobile phase	Solid	Liquid
Imposed	ELECTROPHORESIS	ELECTRO-OSMOSIS
Induced	SEDIMENTATION POTENTIAL	STREAMING POTENTIAL

5

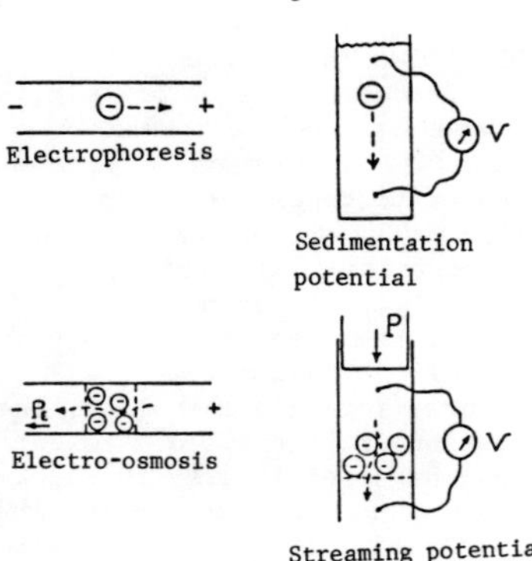

<u>Electro-kinetic effects</u>

2.4 <u>Fundamental laws governing electro-kinetic effects</u>

2.4.1 <u>Electrophoresis</u>

The speed at which a particle in a suspension moves under the influence of an electrical field is governed by the Helmholtz-Smoluchowski formula, which is as follows:

$$u = \frac{\zeta D}{k\eta} \cdot f(\lambda\ \lambda') . E$$

where ζ = the zeta potential of the particles (in mV),
$\quad$ D = the dielectric constant of the suspension liquid,
$\quad$ η = the viscosity of the suspension liquid at ambient temperature,
$\quad$ k = factor to allow for the shape of the particle (k = 4π for a spherical particle),
$\quad$ $f(\lambda,\lambda')$ = function of the electrical resistivity of the liquid and of the particle and its surrounding electrical layer. In practice $f(\lambda,\lambda')$ ranges from 0.75 to 1.5,
$\quad$ E = electrical field applied.

2.4.2 <u>Electro-osmosis</u>

The volume of liquid extracted from a porous layer of immobile solid particles per unit surface area at any time is calculated by the formula established by Yukawa (23), namely:

$$q = k \frac{\zeta D}{\eta} \frac{\varepsilon I}{\lambda}$$

where ε $\quad$ is the porosity of the medium
$\quad$ I $\quad$ is the intensity of the electrical current applied to keep the electrical field at the desired value.

3. APPLICATION TO SPEED UP SEDIMENTATION AND GRAVITY THICKENING OF SLUDGES

Shirato (10) has constructed models of how electrical fields speed up the thickening process, depending on the permeability and compressibility of the sediment. He conducted laboratory experiments on a thick clay suspension in de-ionised water to check the accuracy of his model. He found that the sedimentation rate increases appreciably as the strength of the electrical field applied increases. The model agreed satisfactorily with the laboratory findings for fields of not more than 30 volts per metre. It appears that the properties of the electrical double layer vary, depending on the degree of compaction of the sediment.

Yukawa (8) detected three separate phenomena, one after the other:

(i) first, free sedimentation, at the speed of electrophoretic thickening of the particles plus the speed of gravity thickening;

(ii) second, thickening by consolidation, under the combined impact of gravity compression and of the extraction of water by electro-osmosis;

(iii) finally, compaction under the impact of electro-osmotic pressure alone.

The relative share taken by any one of these phenomena depends on the porosity of the medium at the time. The formula for electro-osmotic pressure is based on the Carman-Kozeny formula and shows that the osmotic pressure is proportional to the electric field applied. The electro-osmotic pressure determines the degree of thickening finally obtained, following a law similar to the law for gravity thickening.

Yukawa (8) quotes the following results for sedimentation tests with and without an electrical field:

Nature of the particles in suspension	Gravity sedimentation rate (in m/s)	Sedimentation rate with an electrical field of 9 volts per metre (in m/s)
Calcium carbonate	$3.5 \cdot 10^{-7}$	$7.6 \cdot 10^{-7}$
Kaolin	$3.2 \cdot 10^{-7}$	$14 \cdot 10^{-7}$
Precipitated silica	$4.2 \cdot 10^{-7}$	$20 \cdot 10^{-7}$
Albumin	$1 \cdot 10^{-7}$	∞

It can therefore be seen that this method raises the sedimentation rate to between two and five times the original value.

The same author also conducted laboratory experiments on a continuous rectangular thickener capable of concentrating a calcium carbonate sludge from 20% dry solid matter to 65% (the same as a filter press operating at 16 kg/cm^2) after just two hours retention time in return for consumption of as little as 14.5 kWh/m^3 on applying a field of five volts per centimetre.

Yukawa continuous thickener (8):

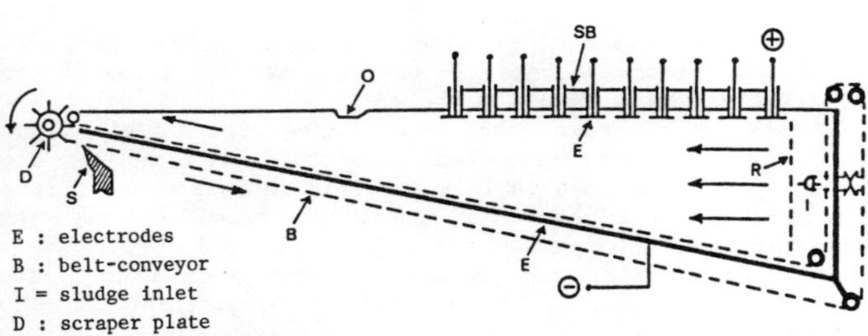

E : electrodes
B : belt-conveyor
I = sludge inlet
D : scraper plate

4. APPLICATION TO CONCENTRATE SUSPENSIONS BY MEANS OF ELECTRO-FILTRATION

Strictly speaking, this is a method of thickening, not dewatering, sludges, since no cake is formed.

The basic principle is continuously to apply an electrical potential of the same polarity as the suspended particulates to the filter surface (or to its metal supports). Once the potential is above a certain critical level, the particles will be sufficiently repelled by the electrode to migrate against the flow of the liquid and, thus, to prevent cake formation (11). The diagram below gives a simplified illustration of the difference between this method and conventional filtration (15):

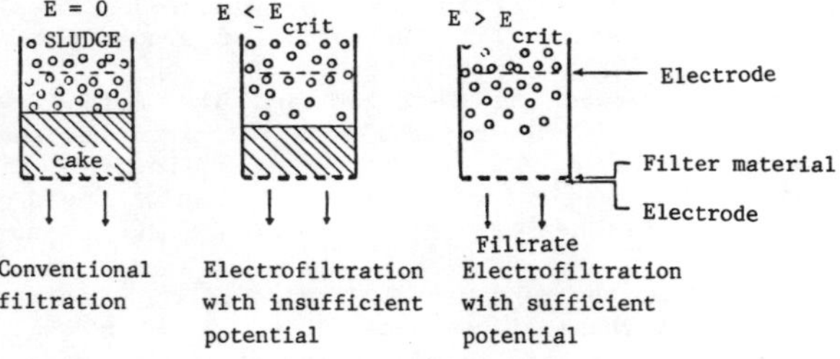

Conventional filtration	Electrofiltration with insufficient potential	Electrofiltration with sufficient potential

The diagram set out below shows the increase in filtrate production over time as a function of the electrical field applied. It shows that the yield can be doubled or even trebled.

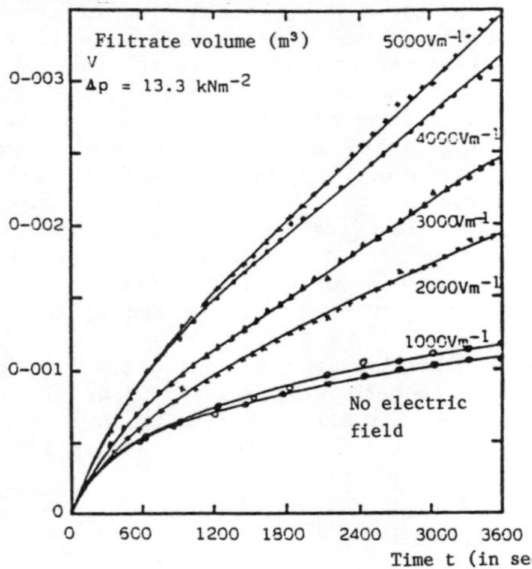

If a cake is formed, an electro-osmotic effect assists in the extraction of water from the cake (at the cathode if, as is generally the case, the particles bear a negative charge).

Moulik (11) has shown that the lines t/v = f(v) plotted for increasingly strong electrical field are straight lines of decreasing steepness (corresponding to a fall in specific resistance to filtration).

Yukawa (8) reached the same conclusion from his experiments on suspensions of calcium carbonate and clay.

Henry et coll. (13) also conducted experiments with a cross-flow system. The speed·of the cross-flow had no significant impact on the filtration rate.

Lee (12) studied a similar method for treating non-aqueous slurries, applying extremely high electrical fields (ranging from 1 000 to 10 000 volts/cm).

5. <u>APPLICATIONS FOR DEWATERING SLUDGES</u>

Sludge can be dewatered by electro-osmosis, which entails placing the sludge between two electrodes and then applying an electrical field between the two to extract water from the sludge. As mentioned earlier, the water is usually drawn to the cathode. Usually any pressure or suction applied to the sludge is kept weak and is intended chiefly simply to draw off the water arriving at the cathode or to keep the electrodes in close contact with the sludge, remembering that the volume of sludge decreases as the operation proceeds.

The difference between this method and the method described in Section 4 is that a cake of dewatered sludge is formed and that the water extracted is not replenished. As a result, it is possible that the cake will not be saturated at the end of the process, which could give rise to specific problems (e.g. higher electrical resistance and losses as a result of the Joule effect).

Yukawa has studied the fundamental factors in the phenomena observed, both with and without mechanical compression (6, 7, 8 and 9). He checked his model against various mineral sludges, including ones containing clays, magnesium hydroxide, silica and calcium carbonate. He emphasised the advantages of this method for dewatering gelatinous sludges, which are easily compressible and therefore difficult to dewater by mechanical compression. The model is based on a calculation of the electro-osmotic flow and takes good account of the quantities involved in the dewatering process. The filtrate yield is proportional to the current density (6). As mentioned earlier (8, 11), the whole process unfolds as if the phenomenon were obeying the law of filtration under constant pressure, with the specific resistance to filtration far lower and electro-osmotic pressure replacing mechanical compression. In addition, the filtrate yield is proportional to the square root of the electricity consumption (7). It is also proportional to the electrical field and to the porosity of the medium, if both are uniform. On the other hand, the ratio between the electro-osmotic flow and current density is unaffected by the strength of the electric field or by the thickness of the bed of sludge to which the field is applied.

By analogy between Darcy's law (expressing the flow rate of a fluid through a porous medium under pressure) and the law governing the electro-osmotic flow rate, it is then possible to introduce the concept of electro-osmotic permeability (K_e) calculated by Casagrande's formula:

$$Q = \frac{K_e \, S \, V}{L}$$

where Q is the electro-osmotic flow;
   S is the surface area of the electrodes;
   V is the potential drop between the electrodes; and
   L is the distance between the electrodes.

Combining Casagrande's formula, which covers the general phenomenon, with the more sophisticated models of electro-osmotic flow constructed by Yukawa or Kobayashi (9), it can be seen that the coefficient of electro-osmotic permeability (K_e) takes account of the viscosity of the fluid, of the porosity of the medium and of the specific surface area of the dispersed solid phase. Consequently, this figure gives an indication of the sludge's propensity for dewatering by electro-osmosis.

As for applications of this method for dewatering sludges, it could be used to treat large volumes of sludge in situ, without any need to transport them. This method has already been used on a large scale to consolidate clay ground, notably by Cambefort and Caron (16, 17 and 18). Both authors found that the flow rate in clays diminished after a few weeks, under the impact of the migration of ions and of the emergence of an area of de-ionised interstitial water, with a particularly high electrical resistance which causes a sharp reduction in the current density across the medium treated. At the same time the pH varies across the equipment, from basic close to the cathode to acid near the anode.

Another option is to treat thin layers of sludge, either in drying beds (19) or in purpose-built machines based on industrial filters for dewatering sludges. The system (patented by Dorr-Oliver Inc.) is one example of such a system (24 and quoted in 14). The unit looks like a rotary vertical classifier. The cake is collected by the submerged part of the vertical disc and then discharged outwards with the aid of a blade. The cathode is a porous sector of the disc coated with a filter medium parallel to the submerged part of the anode. A vacuum is created across the cathode to draw out the water and thereby thicken the cake. Consequently, the unit operates continuously, with the electric field serving not only to repel particles from the filter medium but also to concentrate the cake at the anode. Up to 100 volts can be applied to the electrodes, with a residence time of one minute in the immersion zone.

In France, Ceralit has succeeded to developing a machine producing from a suspension, a strip of clay with a dry matter content of 82% at a productivity rate of 130 kg/m^2/h in return for a total consumption of 0.022 kWh/kg of cake (quoted in 14).

The energy consumption figures given by the various authors vary widely, probably because of the differences in the materials studied, in their initial moisture content and in the length of time allowed for the operation. Deleuil's review of the literature available (14) mentions the following figures:

Researcher	Materials	Energy consumption (in kWh/kg water extracted)
Yukawa	Clays (bentonite and kaolin)	0.05 to 0.1
Von Schwerin	Peat	0.045
US Bureau of Mines	Calcium phosphate	0.021
Deleuil	Clay (kaolinite)	0.56

For comparison, it takes between about 1 and 1.3 kWh to evaporate 1 kg of water.

To a large extent the energy consumption for electro-osmosis depends on the final dry matter content sought. Above a critical dry matter level, the cake degenerates into a very poor conductor, bringing with it a sharp increase in energy consumption as a result of losses due to the Joule effect (6). A balance must be struck between dewatering at low current density, which is slow but has low.energy consumption, and rapid dewatering at high current intensity but with heavy energy consumption.

Few tests have been conducted on sewage sludges from urban sewage plants. Cooling (19) conducted a pilot project treating digested sludges in drying beds and applying an electrical field to some of them but not to others. He found an electro-osmotic permeability constant of 0.006 gal/sq.ft/h per volt/cm. The thinner the layer of sludge, the more economic the process. Cooling's trials demonstrated that the energy consumption was between 0.75 and 1.80 kWh per gallon of water extracted (i.e. between 0.17 and 0.4 kWh per kilogramme of water). This was felt to be uneconomic. Two other problems mentioned were the formation of a dry layer in contact with the anode and the likelihood of corrosion of the anode.

Sunderland (20) compared alternative configurations of electro-osmosis cell:

(i)   cylindrical cell with parallel electrodes to obtain a dry matter content of between 18% and 20% from slurries in return for energy consumption of approximately 0.10 kWh per kilogramme of water extracted;

(ii) immersion cell to capitalise on the hydrostatic charge as well (consumption: between 0.05 and 0.10 kWh per kilogramme of water extracted);

(iii) revolving drum cell (consumption: between 0.11 and 0.13 kWh per kilogramme of water extracted);

(iv) moving belt cell allowing continuous operation and the highest surface/volume ratio. In this case, the energy consumption measured ranged from 0.092 to 0.191 kWh per kilogramme of water extracted, assuming a potential drop of between 10 and 30 volts between the electrodes. The final dry matter content was in the order of 26 to 28%. Although the filtrate has a high pH (over 11) it was of excellent quality.

Scale-up studies (21) have produced a clearer picture of the final design, dimensions and economics of the unit for comparison with other alternatives, as set out in the table below (which is based on 1972 prices):

Method	Final dry matter content	Cost (per tonne dry matter treated)
Centrifuging	15-20%	£20
Vacuum filtration	25-30%	£16
Filter press	35-40%	£20
Band filter press	20-30%	£16-20
Electro-osmosis	26-30%	£13-44 for sludges pre-concentrated to between 2.6 and 10%

6. <u>PROSPECTS FOR APPLYING ELECTRICAL METHODS AND CONCLUSIONS</u>

Closer analysis of the information available shows that:

(i)   At the moment there appear to be no industrial units for thickening sludges by electrical means. The laboratory trials and small pilot projects carried out have been confined to mineral suspensions. There is a danger that the high electrical conductivity of urban sewage sludge could jeopardise the economic prospects of applications of this type.

(ii)   The same applies to electro-filtration as a method of concentration (to form a cake), though some progress is likely (e.g. cross-flow filtration seems promising).

(iii) The chief benefit appears to be for dewatering sludges into a solid residue. Since, however, the cost of the operation depends heavily on the original concentration of the sludge, this process should not be construed as an alternative to mechanical dewatering of slurries but, on the contrary, should be used for extra dewatering of highly concentrated or pre-dewatered sludges (with a dry matter content of over 10% already) upstream of incineration, with its very high energy costs for the water extracted. Either the operation could be carried out intensively on a purpose-built machine (allowing short residence time, treatment of thin layers and strong electrical fields) or less intensively at an intermediate store (with a longer residence time, thick layers treated and a weak electrical field).

The projects now being carried out by Japanese and US research teams bear witness to the revived interest in methods which could be economically promising in specific applications in the medium term.

<u>REFERENCES</u>

(1)   Von HELMHOTZ. (1879). Ann. Physik. Wiedemann. 7, 337.

(2)   STERN, O. (1924). Z. Electrochem. 30, 508.

(3)   GOUY, G. (1910) J. Phys. 9 (4) 457.- Ann. Phys. (1917). 7 (9), 129.

(4)   CHAPMAN, D. (1913). Phil. Mag. 25 (6) 475.

(5)   GRAHAME. (1947) Chem. Rev. 41, 441.

(6)   YUKAWA, H. et coll. (1976). "Fundamental study on electro-osmotic dewatering of sludge at constant electric current". J. Chem. Eng. Japan, Vol.9, No.5, pp.402-407.

(7)   YUKAWA, H. et coll. (1971). "Fundamental study on electro-osmotic filtration. Effect of strength of electric field on flow rate of permeation". J. Chem. Eng. Japan Vol.4, No.4, pp.370-376.

(8)   YUKAWA, H. et coll. (1979)."Studies of electrically enhanced sedimentation, filtration and dewatering processes". Prog. Filt. Sep. pp.83-112.

(9)   KOBAYASHI, K. et coll. (1979). "Electro-osmotic flow through particle beds and electro-osmotic pressure distribution". J. Chem. Eng. Japan Vol.12, No.6, pp.492-494.

(10)   SHIRATO, M. et coll. (1979). "Electroforced sedimentation of thick clay suspensions in consolidation region". AIChE Journ. Vol.25, No.5, pp.855-863, September.

(11)   MOULIK, S.P. et coll. (1967). "Forced flow electrophoretic filtration of clay suspensions. Filtration in an electric field". Journ. Coll. Int. Sci. Vol.24, pp.427-432.

(12)   LEE, C.H. et coll. (1980). "Cross-flow electrofilter for nonaqueous slurries". Ind. Eng. Chem. Fundam. Vol.19, pp.166-175.

(13)   HENRY, J.D. et coll. (1977). "A solid/liquid separation process based on cross-flow and electrofiltration". AIChE Journ. Vol.23, No.6, pp.851-859.

(14) DELEUIL. (1982). "Decantation, filtration et deshumidification des gâteaux sous champ électrique. Approche bibliographique". FILTRA 82, Paris 27-29. October.

(15) WAKEMAN, R.J. (1982). "Effect of solids concentration and pH on electrofiltration". Filtration and Separation Vol.19, No.4, pp.316-319.

(16) CAMBEFORT, H. and CARON, C. (1961). "Electro-osmose et consolidation electrochimique des argiles". Géotechnique Sept. pp.203-223.

(17) CARON, C. (1968). "Applications et essais dans le domaine de l'electro-osmose des terrains sableux et argileux". Bulletin Technique de la Suisse Romande No.1, pp.1-4, January.

(18) CARON, C. (1971). "Consolidation des terrains argileux par electro-osmose". Annales Inst. Techn. Bat. Trav. Pub. No.285, suppt. pp.75-92. September.

(19) COOLING, L.F. (1952). "Dewatering of sewage sludge by electro-osmosis". The water and Sanitary Engineer. November.

(20) SUNDERLAND, J.G. (1976). "Dewatering sewage sludge by electro-osmosis. Part I Basic/studies". Rapport ECRC/N973. August.

(21) SUNDERLAND, J.G. (1977). "Dewatering sewage sludge by electro-osmosis. Part II Scale-up data". Rapport ECRC/N1022. February.

(22) European patent 0 046 155. "Process and apparatus for slime and sludge dewatering".

(23) YUKAWA, H. et coll. (1976). J. Chem. Eng. Japan. Vol.9, p.396.

(24) DORR-OLIVER Inc. (1976). French patent No.77 18568. "Procédé de deshydration de suspensions de matières solides dans un véhicule liquide". US patent No.679 142. June.

DISCUSSION

A.M. Bruce
Do you intend to carry out further work at your Institute on this subject?

F. Colin
No, not at the moment. We carried out some experiments about 15 years ago. These showed that the energy requirements were so great that the process wasn't really economically feasible. The conductivity of the sludge is so high that a lot of energy is required to remove the interstitial water. The dewatering mechanism can really be considered as two separate stages, firstly mechanical and then electrical.

Perhaps the process would have greater application in countries such as Norway where energy costs are much lower?

That is possible.

<u>H.W. Campbell</u>
In the final table, you quote ranges of "final dry matter content". Are these final figures regardless of the initial solids concentration?

Yes, that is the case.

<u>H. Scheltinga</u>
The trouble with electro-osmosis is that it takes a long time, perhaps a few days, to fully stabilise and to prevent odours even from partially stabilised sludges. Can the process be used for fresh sludges?

Yes, it is as effective for fresh as for partially stabilised sludges. The difficulty is, however, the need to treat such thin layers of sludge to avoid the energy requirements increasing dramatically.

INFLUENCE OF POLYELECTROLYTE CHARACTERISTICS
ON SLUDGE CONDITIONING (Lab evaluations)

L. SPINOSA, V. LOTITO, F. LORE' and G. BARILE
C.N.R.-Istituto di Ricerca Sulle Acque,
BARI - Italy

Summary

In recent years polyelectrolytes have found increased utilization
for sewage sludge conditioning before mechanical dewatering.
Molecular weight and charge density are the most important
polyelectrolyte characteristics which affect their effectiveness.
However, little information is found in the literature on which
characteristics polyelectrolytes should have to obtain the most
effective sludge conditioning for each dewatering technique adopted.
Laboratory tests were carried out to study the influence of the
above characteristics on sludge conditioning as a function of the
type of dewatering machine. Results showed that polyelectrolytes
with low molecular weight and high charge density are preferable for
conditioning sludge to be filter-pressed, with medium molecular
weight and high charge density for sludge to be dewatered by belt
press, with medium molecular weight and medium charge density for
sludge to be centrifuged.

1. INTRODUCTION

Dewatering is one of the most important and widely utilized
operations in sludge treatment, but it is both expensive and delicate. It
is usually preceded by chemical conditioning in order to enhance sludge
dewaterability. In recent years polyelectrolytes (as of now indicated by
PE) have found increasing utilization for sludge conditioning even in the
cases (such as pressure filtration) where inorganic chemicals were
traditionally preferred (1). The availability of PE has also allowed new
types of dewatering machines to be developed, such as belt press, whose
performance, in the first step of draining, depends on the phenomenon of
superflocculation which can be accomplished only by organic conditioners.

The Istituto di Ricerca Sulle Acque of C.N.R. (CNR-IRSA) is actively
involved in studies on sludge conditioning and dewatering. In particular,
specific laboratory tests for selecting the type of conditioner,
determining its optimal dosage and predicting the performance of
full-scale dewatering equipment have been developed (2).

In this paper the results of laboratory evaluations carried out to
study the influence of PE molecular weight and charge density on sludge
conditioning, as a function of the type of dewatering machine, are
reported.

2. POLYELECTROLYTE CHARACTERISTICS

PE are natural or synthetic long-chain organic compounds bearing
activated groups which become charged in water solution. Natural PE
include starch and its derivatives, polysaccharides, alginates, etc.;
among the synthetic ones there are polyacrylamides and their copolymers
with acrylic acid and dimethyl aminoethyl acrylate, polyethylene-imines,
polyamines and their condensation products with halogenated compounds,

polyamide-amines, polyethylene-oxides, sulfonated compounds, etc. According to their ionic charge, synthetic PE can be classified as anionic, non-ionic and cationic.

The main characteristics of PE are:
- type, distribution and density of charge
- average molecular weight
- ramification degree
- polymeric chain conformation in solution.

Because sewage sludge particles have a negative electro-kinetic potential, cationic PE are required for their conditioning; the most widely used cationic PE are modified polyacrylamides.

Two models have been proposed to explain PE conditioning mechanism.

According to the first, a destabilizing action of the PE on the electro-negative particles is followed by an aggregation due to inter-particle bridges formation. This mechanism is shown in Figure 1 (reactions I and II); the other reactions (III to VI) do not occur in normal practice and are caused by overdosing or too much shearing of flocculated sludge (3). It is generally assumed that the best effects are obtained when adsorbed PE occupies about half of the particle surface, otherwise a re-stabilization of opposite charge occurs; it appears that charge neutralization is the predominant factor (4).

The second model is more recent and considers the sludge as a semi-rigid matrix with biocolloids or anionic polymers, partly dispersed and partly adsorbed in the matrix; the PE neutralizes biocolloid charges and make possible the sludge dewatering by adsorption or by aggregation (5). It has been suggested that the PE and the natural anionic material contained in biological sludge react to form a gel-like aggregate which enmeshes the larger sludge particles; according to this, the molecular weight appears to play an important role (6).

3. <u>BACKGROUND</u>

Filter press, belt press and centrifuge are the most utilized machines for sewage sludge dewatering. Many laboratory tests are available for characterizing the sludges in relation to their mechanical dewaterability, but little information is found in the literature on which characteristics PE should have for optimum sludge conditioning for each dewatering technique adopted. In general, the flocculation time is a function of the charge density (CD), i.e. low flocculation time is obtained with high CD, while the floc size is dependent on the molecular weight (MW). Filtration of larger size flocs allows higher machine load to be obtained, but the final cake solids concentration results lower.

General criteria for PE selection are given in a WRC Manual (7) together with recommendations about specific laboratory tests.

For filter press a PE with high CD is suggested, while MW is not considered critical, although high MW causes larger flocs and wetter cakes.

In belt press operation two fundamental stages occur: drainage and compression. Rapid and good drainage is obtained if superflocculation of sludge is achieved: PE with high CD are therefore required. PE with high MW are recommended in most cases thus causing a more rapid drainage. Great attention must be addressed to avoid overdosing which could cause belt binding.

For centrifuge a PE with low CD is recommended in order to have a low flocculation rate, thus allowing the PE to continue its action on the sludge flocs after their possible breakdown due to the stresses during acceleration when entering the machine. High MW are preferred for obtaining larger size flocs and therefore high separation rates.

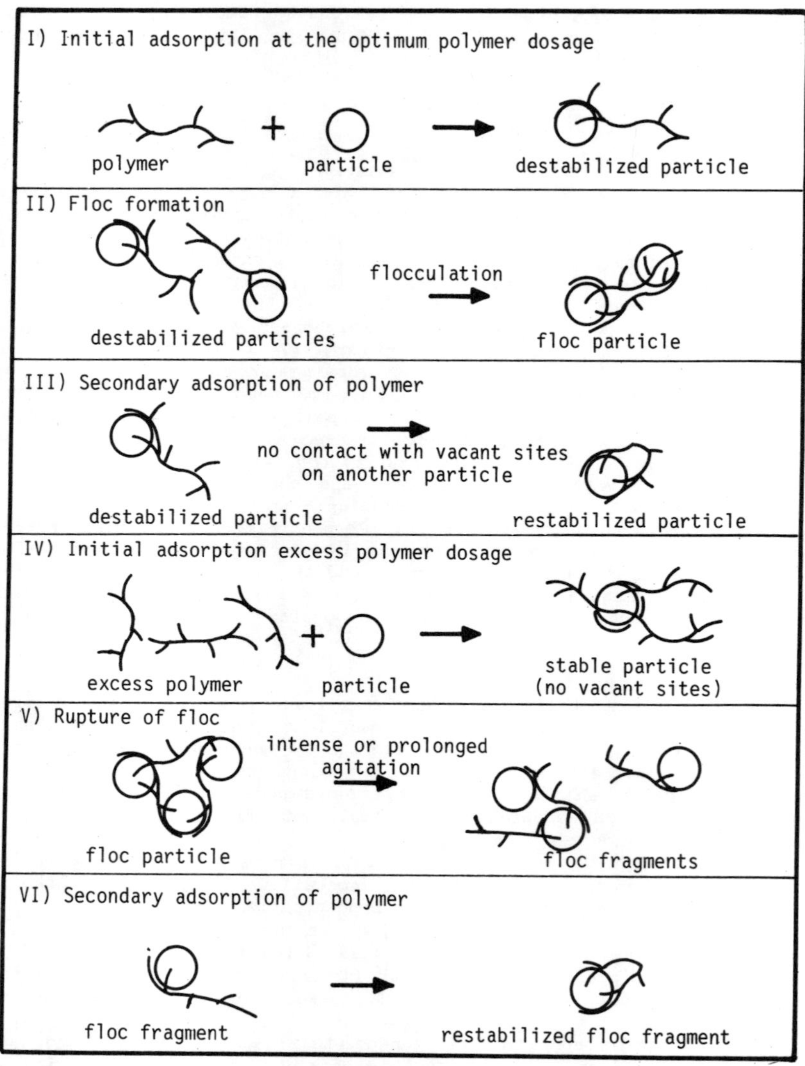

Figure 1. Schematic representation of the bridging model for the destabilization of colloids by polymers (Source 3).

Optimal PE dosage for filter press and centrifuge can be determined by CST measurements after 40 s of stirring at 1 000 rpm in a standard stirrer; for belt press by CST measurements without stirring and a drainage test.

According to other experiences (8), PE with low MW and middle/high CD are suggested for filter press, middle MW and middle/high CD for belt press, middle/high MW and middle CD for centrifuge.

These different conclusions can be explained considering that in the first experience major attention was addressed to the machine load, while in the other to the cake solids concentration.

The above mentioned conclusions are schematically shown in Figure 2.

Considering the scarce information, often in contrast and only qualitative in this field, IRSA started a research programme in order to study the influence of MW and CD of PE on sludge conditioning as a function of the type of dewatering machine. The laboratory tests described in this paper will be followed by pilot scale dewatering tests for a more precise definition of the field of application of different types of PE.

4. EXPERIMENTAL

Municipal sewage sludge was used for tests: characteristics are summarized in Table 1.

Commercially available PE with different MW and CD were used for tests. For MW, data supplied by manufacturers were taken into consideration; they ranged from 2 to 8 million. CD was measured with a method based on the reaction with another PE (9). During this reaction a PE complex is formed, which precipitates near isoelectric point, and an equivalent amount of counter-ions is released in the solution, making possible the indication of the end point by conductometric titration. CD of PE used ranged from 1.4 to 5.2 meq/g. In the following the PE type is symbolized with two letters, the first indicating MW and the second CD, as shown in Table 2.

Tests carried out by IRSA were planned according to two different approaches. In the first, one PE was chosen as standard; the optimal dosages for conditioning sludge to be filter-pressed, belt-pressed, or centrifuged was determined by laboratory tests described in the following, and the other PE were tested at the same dosages as previously determined. In the second the optimal dosages were selected for each PE.

In filtration by filter press the only operating variable which can be affected by PE is the Specific Resistance to Filtration. In these tests values of 1×10^{12} m/kg at 4.9 N/cm^2 were considered as indicating good filterability.

Dewatering by belt press takes place in two basic stages: drainage and compression. In the transition zone it is important to have a cake of such a consistency as to avoid lateral leakage: this can be obtained by adding PE at a dosage able to induce the phenomenon known as superflocculation (10). In this research a methodology consisting of the following steps was used (11):

- CST measurements of conditioned sludge: dosages giving values of about 10-12 s with 10 mm reservoir were considered optimal, because superflocculation occurred;
- gravity drainage of 60 cm^3 of sludge by Buchner funnel apparatus for the time necessary to collect a volume of filtrate amounting to 50% of the initial volume, thus allowing the rate of drainage to be estimated (times higher than five minutes were considered not to be good);
- vacuum filtration of the remaining cake for 10 minutes, thus allowing the final full-scale cake concentration to be predicted.

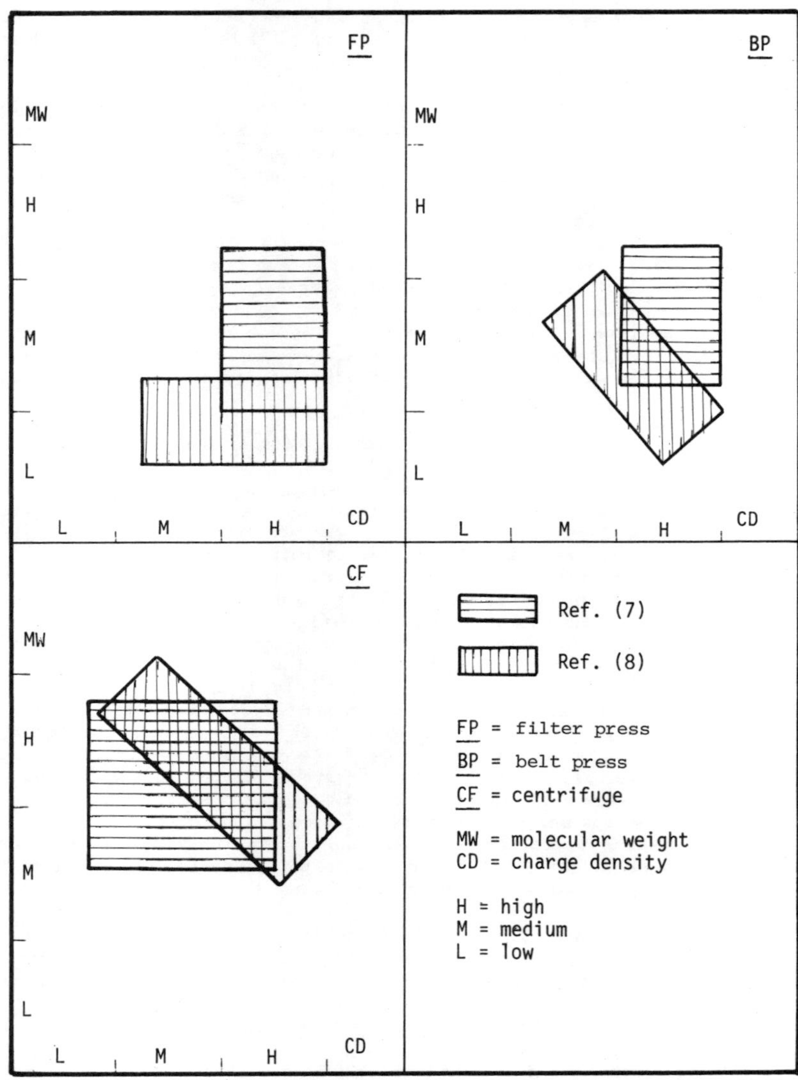

Figure 2. Fields of application of polyelectrolytes.

Table 1. Sludge characteristics.

	Test series					
	1	2	3	4	5	6
Sludge type	An. dig. municipal	Raw mixed municipal	An. dig. municipal	As 3 thickened	An. dig. municipal	As 5 thickened
Initial concentr. (%)	2.23	1.98	1.75	3.82	1.61	4.18
CST Ø18 (s)	100	21	158	152	101	155
Spec. Res. $(m/kg*10^{-12})$	70	19	257	120	92	103

Table 2. Polyelectrolytes characteristics and classification.

Charge Density (meq/g)	Molecular Weight (millions)		
	Low ≈ 2	Medium ≈ 4	High ≈ 8
Low 1.4-2.5	--	--	HL
Medium 2.5-3.5	LM	MM	HM
High > 3.5	LH	MH	--

The dosage to be considered as optimum derives from a compromise among CST value, drainability and cake solids concentration.

A parameter for assessing sludge suitability to centrifugation has not yet been defined because it has not been possible to reproduce on a laboratory scale the conditions occurring in an industrial centrifuge. Centrifugability is seen to be affected by the following sludge characteristics: settleability, scrollability and floc strength. The method based on laboratory scale centrifugation allows both settling and scrolling properties to be evaluated (12), but has not provided satisfactory results with activated sludge due to their poor consistency. For centrifugability evaluation the floc strength method was therefore used (13). It consists in submitting the conditioned sludge to stirring at 1 000 rpm for different times using a standard apparatus and measuring the CST of stirred sludge. The floc strength can be evaluated by plotting CST values vs time of stirring. Good industrial centrifugation results can be expected when:

- the pattern of CST vs stirring is linear with a slight slope between 10 and 100 s;
- the 18 mm reservoir CST value at 10 s stirring is around 10-12 s, or, at least, significative in the whole stirring time range.

5. RESULTS AND DISCUSSION

As far as filter-pressing is concerned, results are summarised in Table 3. PE with high MW appeared to be less preferable than the others. Better results were obtained with low to medium MW and high to medium CD; the combination of medium MW and medium CD does not seem however to give good results. This conclusion was further supported by the results of a few tests carried out measuring the cake solids concentration at the end of a filtration by Buchner funnel apparatus at 4.9 N/cm^2 for 15 minutes: the highest solids concentrations (about 14%) were obtained with LH and MH, followed by MM (12%) and HL (10%).

Results of belt press dewaterability tests are reported in Table 4. PE with high MW seemed to be less effective than the others, especially if associated with low CD. Better results were obtained with medium to low MW and high to medium CD. This conclusion is not so evident in the groups of tests performed at the same dosages for all PE, because sometimes excessive overdosing occurred (for ex. in the test series 4 for MH). Excessive overdosing must be avoided because it causes blinding and drastic cake solids concentration reduction mostly when PE with high MW are used.

As for centrifugability evaluation, the PE characteristics to be preferred are influenced by sludge properties and its history more heavily than filterability evaluation. The most significant floc strength plots are shown in Figure 3. Laboratory tests evidenced that the best floc strengths are obtained with PE having medium to high MW; this can be explained by the fact that solid-liquid separation is not necessarily limited by the floc size considering the high gravity loading, typical of centrifugation. The optimum CD depends on two opposite requirements: the first concerns the availability of PE in solution after the sludge has reached the speed of the centrifuge when entering the machine, thus requiring low CD to have sufficiently slow flocculation; the second is related to the organic strength of the interstitial liquor, thus requiring very high CD for high organic strength liquors, so flocculation can be achieved. Tests showed that medium CD are in general preferable, with a tendency to low or high values according to which of the two described phenomena prevails.

The results obtained are schematically shown in Figure 4.

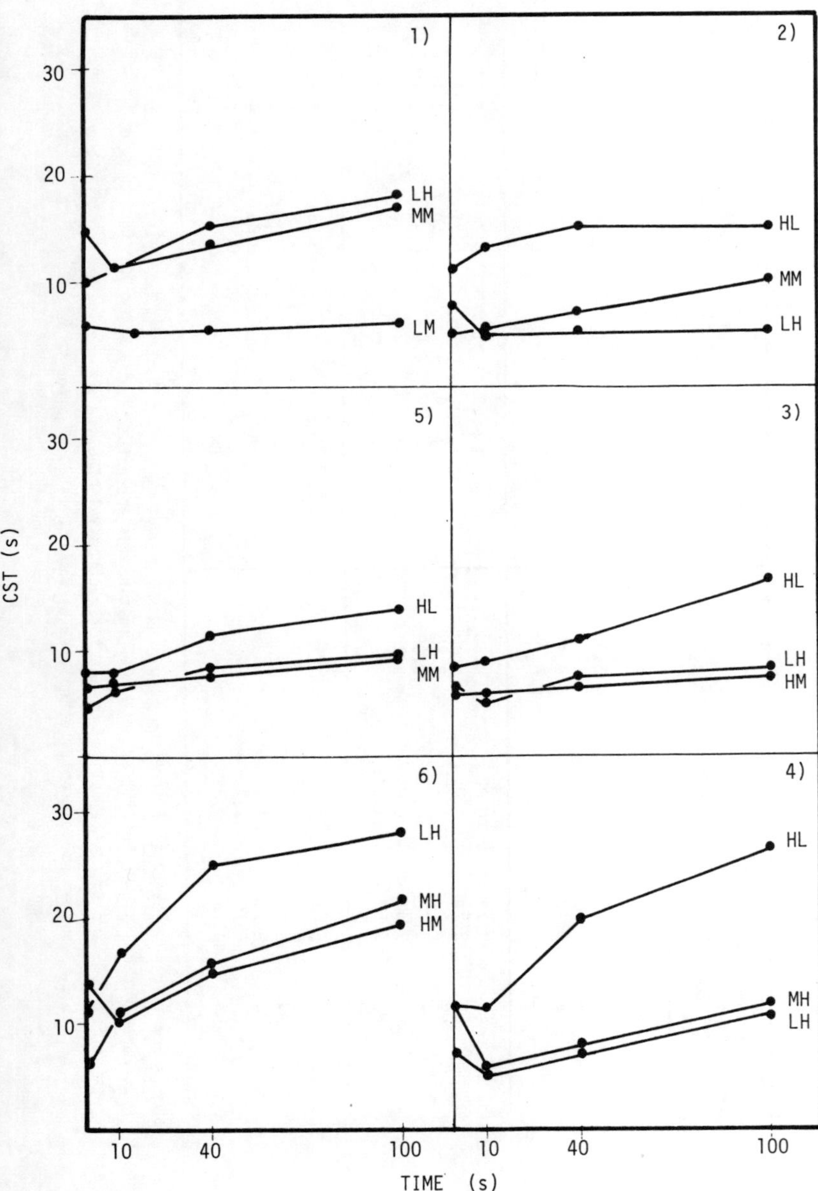

Figure 3. Examples of floc strength plots.

Table 3. Results of tests for filter pressability.

Test series	Polyel. type (MW/CD)	Dosage (mg/kg)	$R_{0.5}$ (m/kg$\star$10^{-12})	R_{CST} (m/kg$\star$10^{-12})	Test series	Polyel. type (MW/CD)	Dosage (mg/kg)	$R_{0.5}$ (m/kg$\star$10^{-12})	R_{CST} (m/kg$\star$10^{-12})
1	LM	6.48	1.05	4.17	2	LM	4.03	0.63	3.43
	LH	4.16	0.90	3.70		LH	"	0.32	2.76
	MM	4.11	1.06	3.60		MM	"	0.60	3.73
	MH	-	-	-		MH	"	0.32	2.40
	HL	5.43	0.53	7.41		HL	"	6.67	8.23
	HM	-	-	-		HM	"	0.65	3.45
5	LM	3.73	0.87	4.86	3	LM	5.71	0.62	3.33
	LH	3.73	1.05	5.35		LH	"	0.77	4.10
	MM	3.73	1.12	3.95		MM	"	1.28	4.06
	MH	3.73	0.87	3.88		MH	"	0.58	3.60
	HL	7.47	0.94	8.81		HL	"	2.29	6.65
	HM	4.98	1.12	-		HM	"	1.03	6.31
6	LM	4.79	1.16	-	4	LM	8.36	0.33	1.78
	LH	4.31	0.64	-		LH	"	0.20	1.35
	MM	6.70	0.60	-		MM	"	0.43	1.59
	MH	5.74	1.34	-		MH	"	0.33	4.45
	HL	6.70	1.29	-		HL	"	0.47	3.36
	HM	4.79	0.89	-		HM	"	-	-

$R_{0.5}$ = Specific resistance to filtration at 0.5 kg/cm^2 vacuum

R_{CST} = Specific resistance to filtration by CST multiradius apparatus

Test series 1, 5 and 6: at the optimal dosage for each polyelectrolyte

Test series 2, 3 and 4: for all polyelectrolytes at the optimal dosage for MM

Table 4. Results of tests for belt pressability.

Test series	Polyel. type (MW/CD)	Dosage (mg/kg)	CST Ø10 (s)	Filtrate vol. at 5' (cm^3) *	Cake conc. (%)	Test series	Polyel. type (MW/CD)	Dosage (mg/kg)	CST Ø10 (s)	Filtrate vol. at 5' (cm^3) *	Cake conc. (%)
1	LM	6.48	12.0	-	18.7	2	LM	6.04	12.6	>30	19.8
	LH	4.70	20.0	-	20.7		LH	"	17.4	>30	21.1
	MM	4.93	20.0	-	21.7		MM	"	15.0	>30	18.4
	MH	-	-	-	-		MH	"	28.0	28	19.8
	HL	2.72	18.5	-	16.0		HL	"	46.0	20	13.7
	HM	-	-	-	-		HM	"	12.4	30	19.8
5	LM	6.22	13.8	>30	21.6	3	LM	7.98	26.0	>30	21.1
	LH	6.22	12.3	5	21.9		LH	"	15.1	>30	21.6
	MM	7.47	17.8	17	22.5		MM	"	15.8	>30	20.9
	MH	4.98	10.8	26	21.5		MH	"	36.5	>30	21.3
	HL	8.71	23.4	>30	21.1		HL	"	24.8	>30	16.8
	HM	4.98	11.7	25	20.9		HM	"	13.8	30	21.8
6	LM	6.70	14.1	24	22.3	4	LM	9.41	43.5	>30	21.6
	LH	5.74	27.9	5	20.2		LH	"	12.8	26	23.9
	MM	6.70	13.2	2	19.8		MM	"	13.3	22	21.6
	MH	5.74	19.3	3	20.6		MH	"	11.9	>30	14.2
	HL	8.62	24.4	3	18.3		HL	"	28.5	>30	12.3
	HM	4.31	23.6	3	18.6		HM	"	99.2	>30	12.6

* Initial volume of sludge to be filtered = 60 cm^3
Test series 1, 5 and 6: at the optimal dosage for each polyelectrolyte
Test series 2, 3 and 4: for all polyelectrolytes at the optimal dosage for MM

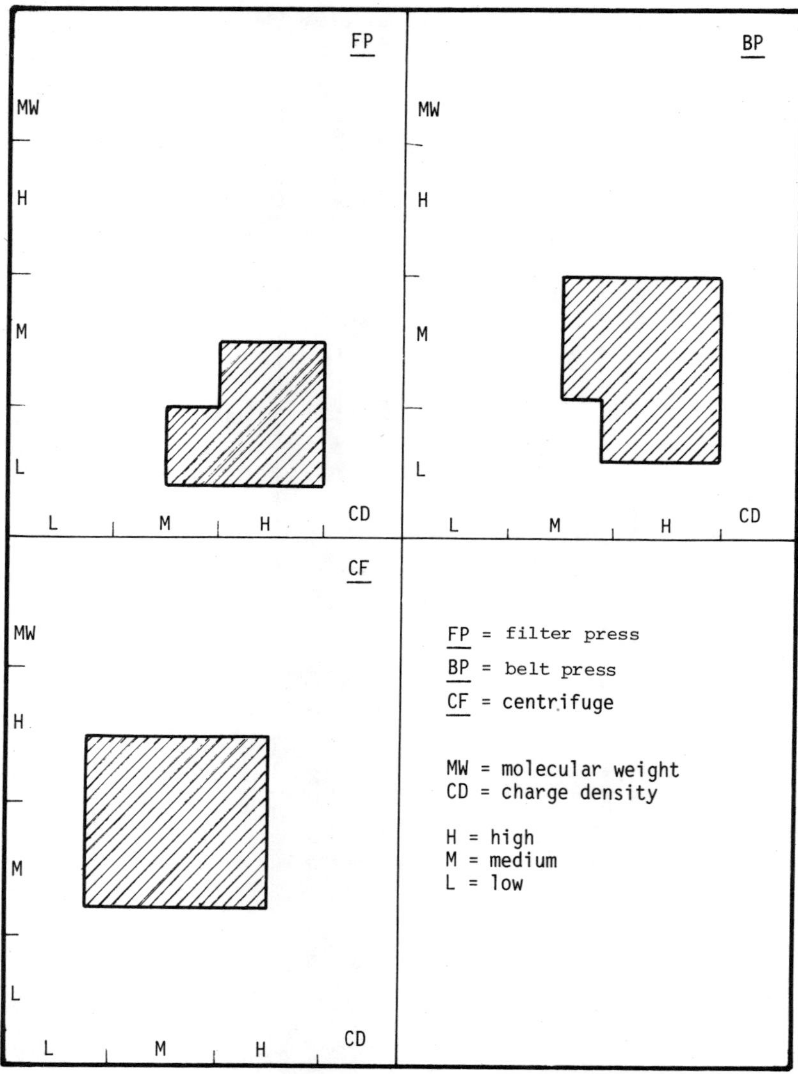

Figure 4. Fields of application of polyelectrolytes (IRSA studies).

6. CONCLUSIONS

Laboratory scale tests were carried out to evaluate how the polyelectrolyte molecular weight and charge density affect the sewage sludge conditioning efficiency as a function of the type of dewatering equipment to be used.

The results obtained allow a first rough selection among the different polyelectrolyte types available on the market to be done, thus limiting the number of tests for selecting the best type for a certain application.

Pilot scale dewatering tests will follow in order to define more precisely the field of application of different types of polyelectrolytes.

REFERENCES

(1)   BRUCE, A.M., CAMPBELL, H.W. and BALMER, P. (1984) Developments and trends in sludge processing techniques. Proc. of the EEC 3rd Int. Symp. on "Processing and use of sewage sludge", D. Reidel Publ. Comp., 19.

(2)   ISTITUTO DI RICERCA SULLE ACQUE (1984). Metodi analitici per i fanghi: parametri tecnologici. IRSA Report 64.

(3)   U.S.EPA (1979). Process design manual for sludge treatment and disposal. Report 625/1-79-011.

(4)   GREGORY, I. (1983). Sludge characteristics and behaviour. NATO ASI Series E-66, Martinus Nijhoff Publ., 1.

(5)   BERTACCHI, B., SECONDINI, I. and ALBANESE, P. (1983). Criteri di scelta e metodi di valutazione di prodotti condizionanti per la disidratazione dei fanghi. Proc. of the Seminar on Sludge mechanical dewatering, IRSA Report 65, 35.

(6)   ROBERTS, K. (1978). The scientific basis of flocculation. Sijthoff and Noordhoff Publ., 3.

(7)   LOCKYEAR, C.F., JACKSON, P.J. and WARDEN, J.H. (1983). Polyelectrolyte Users' Manual. Water Research Centre Technical Report, TR 184.

(8)   BERTACCHI, B. (AUSIND, Montedison) (1983). Private communication.

(9)   TIRAVANTI, G., LORE', F. and SONNANTE, G. (1985). Influence of the charge density of cationic polyelectrolytes on sludge conditioning. Water Research, 19, 1, 93.

(10)   BASKERVILLE, R.C., BRUCE, A.M. and DAY, M.C. (1978). Laboratory techniques for predicting and evaluating the performance of a filterbelt press. Filtration & Separation, 15, 5, 445.

(11)   SPINOSA, L., MININNI, G., BARILE, G. and LORE', F. (1984). Study of beltpress operation for sludge dewatering. Proc. of the Conf. on "Solids/liquids separation practice and the influence of new techniques", Leeds, 2-5 Apr., 85.

(12)   VESILIND, P.A. (1970). Estimation of sludge centrifuge performance. J. Sanit. Engin. Div., ASCE, 96 (SA3), 805.

(13)   SPINOSA, L. and MININNI, G. (1984). Assessment of sludge centrifugability. Proc. of the EEC Workshop on "Methods of characterization of sewage sludge", D. Reidel Publ. Comp., 16.

DISCUSSION

H.W. Campbell
Is there a simple correlation between the characteristics of the polyelectrolyte and the cost? Certainly, some are so expensive

L. Spinosa
There doesn't appear to be any simple correlation. However, it probably is too early to be certain since with many polyelectrolytes

that one can't really afford to select them for routine usage.

only laboratory scale trials have been carried out. It is certainly true that a number of poly-electrolytes are very expensive, for example, those with a high charge density.

B. Paulsrud

You have restricted your survey to cationic polyelectrolytes. However, anionic polyelectrolytes also are used in Norway. Can you apply your results to these types of polyelectrolytes?

I have examined only anionic polyelectrolytes and so I can't really comment.

J.G. ten Wolde

Is the concentration of polyelec-trolyte used related to the total solids content of the sewage sludge sample or to the type of solid?

It is related to the total solids present.

F. Colin

Should other parameters be taken into account?

No, these are not really necessary.

A.M. Bruce

Is the floc strength test good enough to predict the concentration of polyelectrolyte required to efficiently dewater sludge using other tests such as scrollability?

We must have carried out some 200-300 pilot plant studies and these suggest that the floc strength test is adequate for this purpose.

What centrifuge did you use?

A Sharples P600, treating between 600-700 cm^3 and 1 000 cm^3 samples.

H.W. Campbell

From the floc strength test you are able to indicate the appropriate polyelectrolyte type that should be used. Surely, this won't predict the concentration of the final solids?

I agree that it is not always possible to predict the final solids concentration. It is interest-ing, however, to note just how good was the correlation between the results from pilot and full-scale experiments.

THE CHP-FILTER PRESS - THE FIRST CONTINUOUS HIGH-PRESSURE FILTER PRESS

U. LOLL
Abwasser - Abfall - Aquatechnik
D-6100 DARMSTADT, FRG

Summary

With the invention of the CHP-filter press the development of a continuous working high pressure press has been succeeded by the first time at which the pressure on a suspension to be dewatered can be continuously increased and therefore dewatering results which were so far only able to be achieved with chamber-, plate- or membrane- filter presses are obtained. The development character-istics, the constructive realization and the field of application as well as the efficiency of the equipment are described below.

1. CHARACTERISTICS OF THE CONVENTIONAL MACHINES FOR THE SOLID/ LIQUID SEPARATION OF SLUDGE-LIKE SUSPENSIONS

The use of the so far existing machinery for solid/liquid separation, such as centrifuges, decanters, vacuum filters, screen and belt presses, as well as chamber-, plate- and membrane-filter presses include the following problems. In the case of a reasonable amount of machinery and scheduled continuous operation, in many cases of application the dewatering results are insufficient. In order to obtain satisfactory dewatering results, a considerable amount of equipment, and therefore high investment costs, as well as a non-continuously working procedure, had to be accepted.

The following characteristics are particularly relevant:

1.1 Centrifuges and decanters

Continuous working method, favourable relation between machinery and flow rate, high capacity, low dewatering results, bad separation cut. The separation cut and the dewatering results differ greatly according to the specific weights in the sludge liquor. With increasing specific weight difference, the dewatering results improve.

1.2 Vacuum filters

Continuous working method, unfavourable relation between machinery and flow rate, high capacity, low dewatering results, good separation cut, partial vacuum up to approx. 0.6 bar.

Even the addition of press-belts does not show significant improvement in the dewatering results.

1.3 Screen- and belt-filter presses

Continuous working method, reasonable relation between machinery and flow rate, small capacity, reasonable dewatering results, reasonable separation cut, short-time pressure of up to approx. 2.5 kp/cm^2.

Since the pressure is too low and the time of activity is too short, the real dewatering results can only be insignificantly improved.

1.4 Chamber-, plate- and membrane-filter presses

Non-continuous working method, very unfavourable relation between machinery dimensions and flow rate, high capacity, good dewatering result, good separation cut, long-term filtration pressure up to 20 bar, respectively 15 kp/cm^2 of membrane filter presses.

2. DEVELOPMENT CHARACTERISTICS AND DEWATERING PROCEDURE OF THE NEW CONTINUOUS WORKING HIGH PRESSURE FILTER PRESS

For the new development of the described continuous working high pressure filter press, the following requirements in technology, efficiency and optimal machinery were postulated:
- continuous sludge feeding
- continuous increase of pressure
- continuous cake outlet
- continuous filter cloth washing
- highest possible regulation of the flow rate
- good dewatering results
- high separation cut
- low amount of machinery (small dimensions)
- low energy demand
- highest possible pressure up to 20 kp/cm^2
- pressure control in amount and activity time.

The dewatering procedure of the high pressure filter press, of which the most relevant elements are shown in Figure 1, runs as follows:
- The suspension is fed continuously into the machine by a helical rotor pump via a static pre-dewatering-chamber stage V (which is not shown in Figure 1 due to lack of space).
- A closed sludge chamber (stage K, R and H) changes its volume according to the dewatering progress).

The sludge chamber consists of two endless filter belts, two endless channel belts, two endless asided seals, and two endless armoured belts consisting of high pressure plate elements.

All sludge chamber elements are in operation synchronously within the machine. The dewatering in the closed chamber is effected by successive following pressure phases, being individually controllable in amount and activity period. Amount and kind of reaction of the individually chosen pressure always correspond to the specific dewaterability of the suspension to be dewatered.

The following pressure phases were passed through:
- Increasing hydrostatic filling and filtration pressure up to 0.25 bar (stage V) turning into an
- increasing static pressure up to 0.5 kp/cm^2 (stage K), turning into an
- increasing dynamic pressure up to 2.5 kp/cm^2 (stage R), turning into an
- increasing static pressure up to 20 kp/cm^2 (stage H).

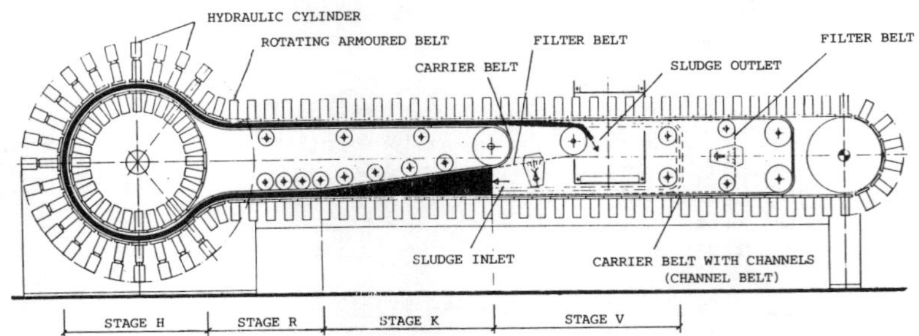

Figure 1. Elements of the continuous working high pressure filter press.

Figure 2. The CHP-filter press in operation on a sewage treatment plant.

3. <u>CONSTRUCTIVE REALIZATION OF THE REQUIRED PRESSURE REACTIONS</u>

The hydrostatic filtration pressure up to 0.25 bar is produced by gravity of the filling height of the suspension itself. Within a 0.6 m Ø filter chamber consisting of motionless filter cloth the filling height up to 2.5 m can be controlled. The filter efficiency of the chamber is activated and increased by two speed-controlled wipers.

By steady reduction of the inflow and increasing solid matter concentration, the hydrostatic filtration pressure turns into an increasing static pressure up to 0.5 kp/cm^2. This pressure is produced by the progressive sphenoidal reduction of the chamber channel belts, a number of pressure rolls, and the two side walls. Pressure rolls and side walls are built in, whilst the filter and channel belts move synchronously with the suspension, as controlled. The turning into the dynamic pressure is achieved infinitely.

The controlled dynamic pressure up to 2.5 kp/cm^2 is produced by increasing pressure of stationary built-in pressure rolls. All chamber elements, like the two filter belts, the two channel belts, and the two asided seals, operate synchronously. The chamber volume is automatically reduced with the progress of dewatering.

The increasingly controlled static pressure up to 20 kp/cm^2 is produced by plate elements with built-in hydraulic cylinders. All chamber elements including the pressure producing plate elements with built-in hydraulic cylinders, co-operate smoothly and synchronously. The chamber volume is also reduced automatically with the progress of dewatering. According to building dimensions of the machine the total static pressure produced at the same time is up to 1 500 Mp.

Since all pressure transmitting elements do not move in relation to each other, but only mutually and synchronously with the medium included, a smooth pressure transmission to the dewatering chamber and the suspension is possible. The pressure power only reacts within the pressure elements. Therefore outside effecting pressure and weight is not produced, and the foundation has to carry only the weight of the machine itself.

In Figures 3 and 4 some important details of the construction are shown.

Figure 3. Hydraulic system.

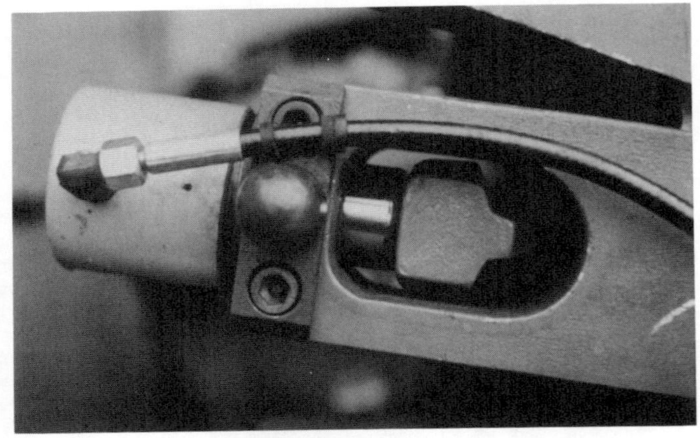

Figure 4. Single hydraulic cylinder.

4. SPECIFIC MACHINE DATA AND EFFICIENCY RANGE OF THE VARIOUS MODELS

The new continuous working high pressure filter press is built in six different basic types according to various applications. The most important technical data for the machine and the efficiency range of the various models are compiled in Tables 1-3. Special constructions are possible for individual applications where certain phases of dewatering have to be extended resp. shortened in order to receive an optimal dewatering result for problematic cases.

Table 1. Technical machine data.

type	filter cloth width m	cylinder diameter m	length m	width m	width with stage V	hight m	hight with stage V	total power capacity kW	total weight Mp
H1/0,5/1,2	0,5	1,2	7,5	1,2	2,5	2,0	2,4	4,5	~ 5
H2/1,0/1,2	1,0	1,2	7,5	1,7	3,0	2,0	2,4	6,8	~10
H3/1,5/1,2	1,5	1,2	7,5	2,2	3,5	2,0	2,4	7,8	~20
H4/1,5/1,5	1,5	1,5	8,5	2,2	3,5	2,3	2,9	7,8	~25
H5/1,5/1,8	1,5	1,8	9,5	2,2	3,5	2,6	3,4	9,2	~30
H6/1,5/2,1	1,5	2,1	10,5	2,2	3,5	2,9	3,9	9,2	~35

Table 2. Pressure and time of active pressure.

type	stage V pressure amount bar	stage V pressure time min	stage K pressure amount kp/m²	stage K pressure time min	stage R pressure amount kp/m²	stage R pressure time min	stage H pressure amount kp/m²	stage H pressure time min	total time of active pressure min
all types	0,05 to 0,25	1,5 to 15	up to 0,5	1 to 7	up to 2,5	1 to 7	up to 20	2,2 to 15	5,7 to 44

Table 3. Active static and dynamic filter cloth values - efficiency range of different machine types.

Type	active static filter cloth values total- -length m	active static filter cloth values total- -area m²	active dynamic filter cloth value length m/h	active dynamic filter cloth value area m²/h	active dynamic filter cloth value supporting speed m/sec	cake thickness mm	cake production m³/h	sludge inflow m³/h
H1/ 0,5/1,2	5,9 to 7,9	6,0 to 9,8	11 to 77	10 to 70	0,003 to 0,021	15 to 30	0,07 to 1,0	up to 10
H2/ 1,0/1,2	5,9 to 7,9	11,6 to 15,4	11 to 77	21 to 147	0,003 to 0,021	15 to 30	0,15 to 2,1	up to 20
H3/ 1,5/1,2	5,9 to 7,9	16,9 to 20,7	11 to 77	32 to 224	0,003 to 0,021	15 to 30	0,25 to 3,3	up to 40
H4/ 1,5/1,5	7,5 to 9,5	21,8 to 25,6	14 to 100	40 to 280	0,003 to 0,021	15 to 30	0,3 to 4,2	up to 60
H5/ 1,5/1,8	9,1 to 11,1	25,8 to 29,6	18 to 126	52 to 364	0,003 to 0,021	15 to 30	0,4 to 5,6	up to 80
H6/ 1,5/1,2	10,7 to 12,7	30,7 to 34,5	22 to 150	63 to 440	0,003 to 0,021	15 to 30	0,5 to 6,7	up to 100

All types can be equipped on request with a continuous operating sludge washer, which might be installed for the dewatering of certain sludges of mostly industrial sources.

5. FIELD OF APPLICATION AND CAPABILITY OF THE AGGREGATE

5.1 Field of application
The new continuous working high pressure filter press is suitable for the dewatering of sewage sludges, suspensions of industrial production and sediment sludges.

Since the amount as well as the time of pressure activity can be controlled in a wide range, the whole scale of aimed dewatering degrees can be covered. If requested, a conditioning of the material with organic as well as with inorganic coagulants resp. flocculants or thermal conditioning is possible. With the new kind of sewage sludge dewatering it is possible to produce a final product for land fill disposal or grain filter cake for agricultural use with one and the same machine.

The compact construction and the low weight of the machine make it suitable for many types of implements, where conventional dewatering machines can no longer meet the increased flow rate demand resp. the higher solid content of the filter cake.

5.2 Efficiency
Based on the high and long lasting active pressures, the new press achieved final solid matter qualities in the filter cake under all applications tested so far, i.e. from known chamber-, plate- and membrane-filter presses and especially in direct efficiency comparison with decanters and belt filter presses. The possible dewatering results in connection with the conditioning with organic flocculation aid are clearly exceeded. Since sewage sludge solid matter results ranged betwen 35-55% dry matter content industrial suspensions were partly dewatered up to more than 75% dry matter continuously.

6. PARALLEL TEST RESULTS WITH CENTRIFUGE, BELT FILTER PRESS, AND CHP-FILTER PRESS
In order to judge the efficiency of the new CHP-filter press, parallel test results under technical conditions with centrifuge, belt filter press and CHP-filter press are shown in the following:

Based on an anaerobic stabilized sewage sludge of a municipal water treatment plant with approx. 6.4% dry solids content, and after thickening by gravity, a sludge conditioning with an organic polyelectrolyte (BASF CF 600) was performed before the dewatering.

The quantity of polymers used for all three dewatering processes was kept in the same range in order to have absolutely equal basic conditions for the three machines. The individual dewatering results for the dewatering presses are compiled in Figure 5.

During the test of the CHP-filter press, two different final pressure stages in the system were applied. At a final pressure in the H-phase of 5 bar for about 4 minutes, solid matter results between 34 and 38% in the filter cake were reached.

At an increased final pressure of 15 bar for about 8 minutes a clear improvement of the results, namely between 40 and 44.5% solid matter was achieved. When comparing these results achieved by the technical tests with those achieved by Perchthaler and Stefou for the dewatering of sewage sludge with high pressure zones of a belt filter press, (see Figure 6) a principle conformity in results is achieved.

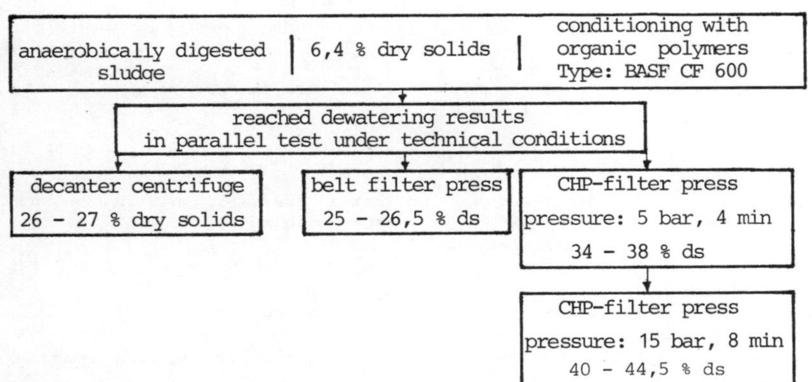

Figure 5. Results of a parallel test with three different dewatering
machines with special reference to the CHP high pressure filter
press.

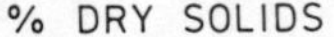

% DRY SOLIDS

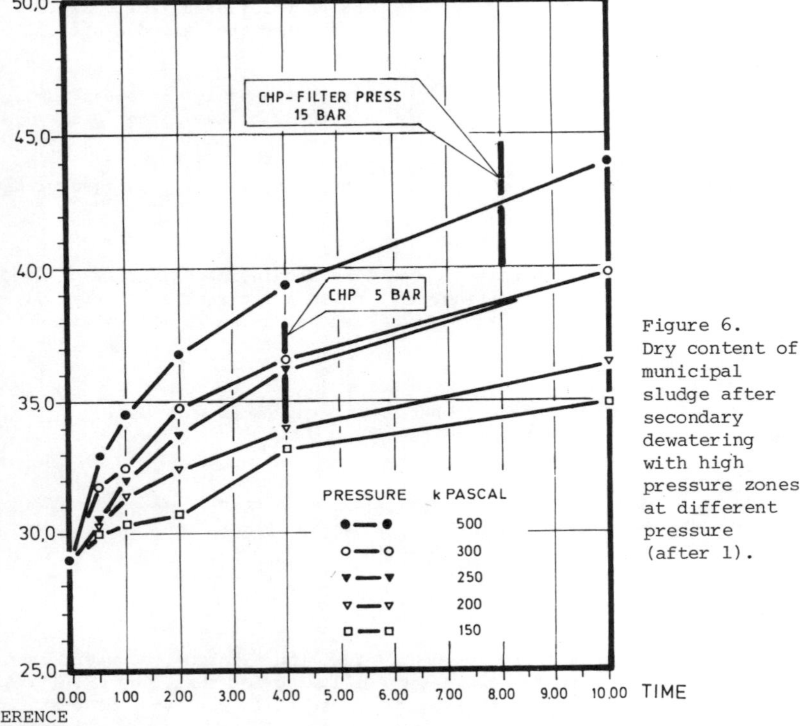

Figure 6.
Dry content of
municipal
sludge after
secondary
dewatering
with high
pressure zones
at different
pressure
(after 1).

REFERENCE

(1)   PERCHTHALER, H. and STEFOU, St. (1984). Filter belt presses with
high-pressure zones. Recycling International, EF-Verlag Berlin,
416-421.

DISCUSSION

A. Wellinger
How do the power costs of the CHP-filter press compare with those for other presses?

U. Loll
It is much cheaper to run.

P.J.W. ten Have
It is important to consider the kWh/m^3 sludge treated for each process to get comparable results.

That is the basis on which we have made our comparison.

L.P. Savelkoul
Is the CHP-filter press still cheaper if you consider all costs involved, for example, manpower, capital and energy?

It costs about the same in total as a belt press but, of course, the final results are so much better.

P.J.W. ten Have
Is the CHP still being developed?

Yes, as with all new machines, it is still being developed and licences for its exploitation are being considered, for example, Japan and the United States.

H.W. Campbell
Is the pre-thickener really necessary?

Yes, it greatly increases the capacity of the filter press.

P. Ockier
How long does it take to change the belts?

They can be easily changed within 2 hours.

M. Santori
Is there a problem with noise whilst the machine is operating?

The machine makes little or no noise as it runs so slowly. The loudest noise is the belt washing unit.

A. Wellinger
Are there problems with the rubber sealants at the edges of the belts?

No, they work very well.

T.J. Casey
There is no relative movement between the belts. However, you only show one drive system for the belts. Surely you must get slippage between the belts?

No, each belt is independently driven.

L.P. Savelkoul
It is important to compare the best results obtained with any press and not only those obtained with the same polyelectrolyte.

Yes, that is true. However, even allowing for that, the CHP-filter press still performs the best.

P. Balmer
What is the longest period of
continuous operation that you have
experienced with the CHP-filter
press?

14 days.

P. Ockier
Was a pre-thickener used?

Yes and decanters were also
installed.

A.M. Bruce
It is important to note that large
throughputs can be achieved with
other presses which are not
operated continuously.

That is true. However, you have to
wait until the end of the batch
cycle to check on the suitability
of the conditions used. With the
CHP-filter press and continuous
operation these can be checked
after only a few minutes and so
control is very much better.

Can the filter press be fitted into
a smaller building than with other
filter presses?

Yes and so the capital costs will
be that much lower.

When can we see these plants in
operation?

There are two technical plants in
the Federal Republic which are in
operation now. However, the production model will not be available
until this Autumn. It is hoped that
licences to build the plants will
be issued at about that time. It is
intended to produce the plant in a
number of different sizes.

A DUTCH APPROACH TO MANURE PROCESSING

P.J.W. ten HAVE
Government Agricultural Waste Water Service (R.A.A.D.)
ARNHEM, The Netherlands

Summary

Within a short time laws will be introduced which will allow maxima
to be set on the amounts of fertilizer applied to agricultural land.
This will result in a sharp increase in the transport of animal
slurries to farms with a shortage of minerals. Regarding the small
demand for veal calf manure most of it will have to be purified.
Plans to do this and the results of a central plant at Elspeet are
discussed. For the longer term future it is expected that surpluses
of pig manure will have to be treated as well. Two possible processes
are discussed. Attention is paid to the problem of the ultimate
disposal of the residual products.

1. INTRODUCTION

The last twenty years has shown a tremendous growth in the Dutch
production of pig and poultry meat. This has been accompanied by an
equivalent amount of animal excrements, which, until now, has been spread
on agricultural land. In many cases the doses applied exceeded the amount
of minerals needed for plant production. One of the results is an
increasing nitrate level in groundwater used for the production of drinking
water. Over-dosages of phosphorus threaten, in the long run, the health of
surface waters (eutrophication), whereas plant quality and soil
productivity can be harmed by repeated excess application of metals such as
cadmium and copper.

Very recently, a detailed study was published in which the manure
surplus of every individual Dutch farm with animal production was
calculated (1). Based on a standard which only takes plant productivity
into account (I.B.-standard) the total surplus of all individual farms in
1982 appeared to be no less than 18 million m^3 per year. If not more than
70 kg P_2O_5/ha/year is allowed, the annual surplus increased to about
40 million m^3, which was about half of the total Dutch animal slurry
production (86 million m^3) in that year.

The pollution of soil and groundwater by animal excrements will
gradually be diminished in the very near future, however. The first of a
number of laws for the protection of the soil was implemented in January
1985. By January 1986 the introduction of these laws is expected to be
completed. They give the possibility to put a maximum on the annual
fertilizer dosage and charge the manure banks with the disposal of the
surplus that results from the imposition of such a standard. At this moment
it is, however, far from clear what the final solution of the problem will
be.

In the next chapters the directions in which the solutions are sought
will be discussed with special regard to processing.

2. WHAT TO DO WITH SURPLUSES

Figure 1 gives a general outline of the possibilities for manure
disposal. The main flow is expected to go, as usual, to arable land
belonging to farms with little or no animal production. (There is no need
for extra manure on grassland). It is not known, however, how much animal

manure in the years to come will be accepted by farms with a mineral shortage. Environmental constraints will, undoubtedly, limit the acceptability of animal manure for the farmers who, it should not be overlooked, are free to choose between organic and artificial fertilizers. Whatever the acceptability will be, one can expect that in a few years a larger volume of manure will be transported from one farm to another over an increased distance. The activities of the manure banks will grow dramatically in order to guide the transport of the surpluses to areas with a mineral shortage.

Depending on the standards for maximum manure dosages and the market for animal slurries, other routes will have to be used. Figure 1 shows that processing is involved in all these routes. It is obvious that the demand for slurries with low concentrations of dry matter and plant nutrients like manure of veal calves (1-2% dm), sows (1-6% dm) and pigs (6-8% dm) will only be minimal, so if surpluses have to be processed it will undoubtedly be this category. Whatever the type of manure or process, separated material still remains to be disposed of. In the following chapters this aspect will be elucidated.

3. VEAL CALF MANURE

Since 1976 veal calf manure has been purified in a central treatment plant at Elspeet. Figure 2 shows the layout. Three mornings per week tankers collect slurry in the area, transport it to the plant and discharge their load into a reception pit. After coarse screening the manure is transferred into a balancing basin. Every morning the daily volume of manure to be treated is pumped into the aeration basin. This takes about an hour. Shortly before that moment the aerators are started. About twenty hours later the aeration is stopped in order to separate the purified liquid from the activated sludge. After approximately one hour sedimentation the effluent is discharged to the sewer system of the village. At the end of the aeration period surplus sludge is pumped to a sedimentation annex balancing basin. This sludge is periodically transported by neighbouring farmers and used as fertilizer on arable land.

The system is discontinuous in order to maximize denitrification in a substrate with a low COD/N ratio (ca. 5). This is elucidated in Figure 3 which gives a picture of the level of several parameters during an aeration period in the aeration tank after separation of the suspended solids. This was performed by laboratory centrifugation directly after the samples were taken. Immediately after the supply of a fresh manure dosage the potential oxygen consumption roughly equals and often exceeds the oxygenation capacity of the aerators. The resulting low oxygen level causes a simultaneous nitrification and denitrification until, due to the oxidation of the easily biodegradable material, the oxygen level starts to rise and denitrification stops. Most of the BOD is absorbed in about an hour after the manure supply; at the end of the cycle about 10 ppm is all that remains. After sedimentation the BOD concentration of the effluent is generally higher, due to suspended solids which can be removed by centrifugation but not by sedimentation in an hour. More information about the performance of this process can be found in Table 1 which shows the average influent and effluent concentrations in the period March 1984 to January 1985. The surplus sludge production is about 0.1 m^3 per m^3 manure. This sludge contains about 4-5% dm. Neighbouring farmers come to collect it, mostly at their own costs, whereas they have shown no willingness at all to accept veal calf manure itself.

The activity of the activated sludge, measured as the rate of nitrification, increases about proportionally with temperature. Due to the fact that two out of three aerators are surface aerators and that the

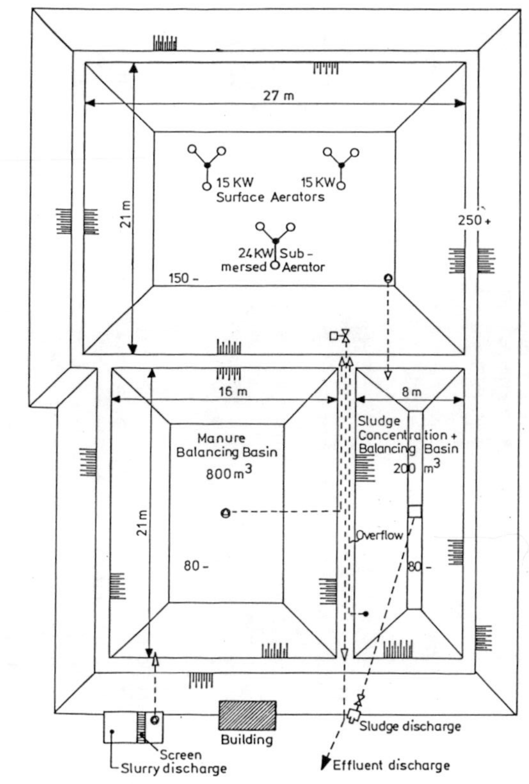

Figure 1. Routes for the disposal of animal slurry and residual products from processing.

Figure 2. Layout of the central purification plant for veal calf manure at Elspeet.

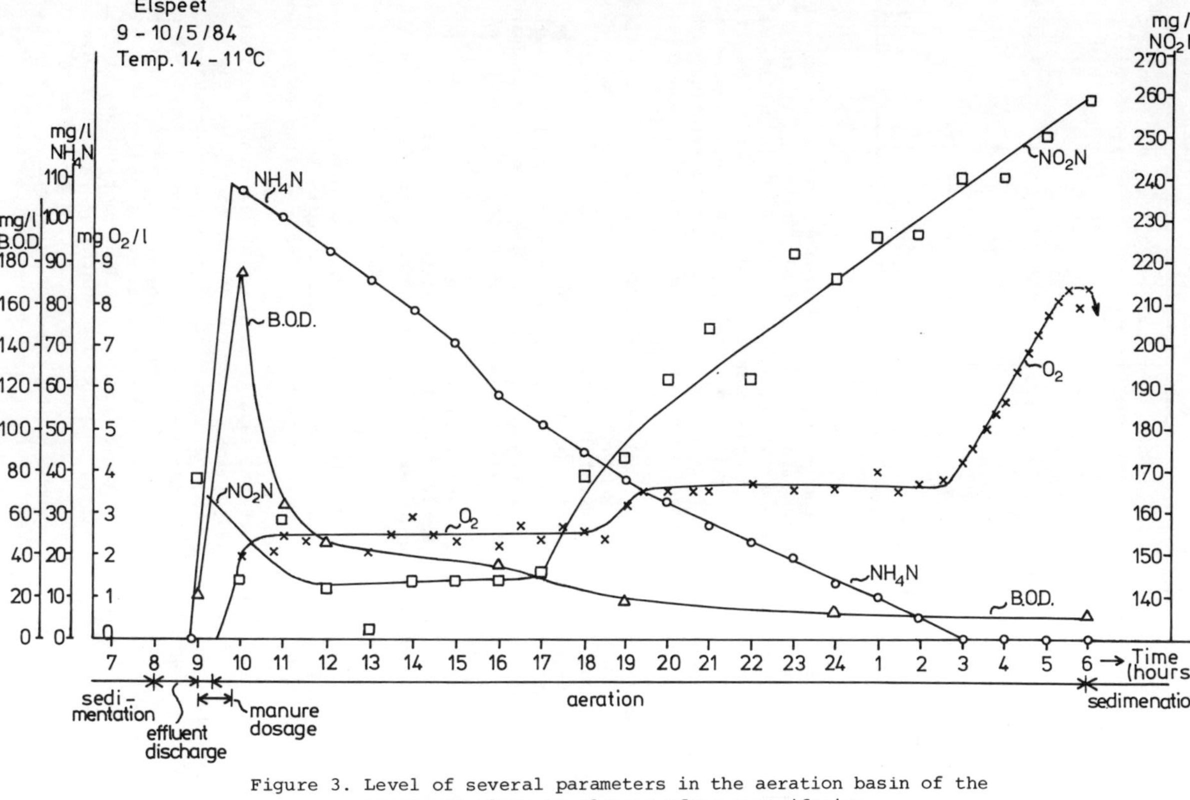

Figure 3. Level of several parameters in the aeration basin of the treatment plant at Elspeet after centrifuging.

Table 1. Composition of influent and effluent of the purification plant for
veal calf manure at Elspeet. Period March 1984-January 1985.

Parameter		Influent n	Effluent n	Reduction %
COD	mg/l	12 500 (8)	1 370 (43)	89.0
BOD	"	5 700 (2)	79 (42)	98.6
Nkj	"	2 500 (7)	74 (43)	97.0
NH_4 -N	"	2 000 (3)	18 (43)	99.1
NO_2 -N	"	--	147 (42)	
NO_3 -N	"	--	4 (41)	
N total	"	2 500	225	91.0
P	"	350 (6)	357 (3)	
Dry matter	"	13 400 (7)	--	
Ash	"	8 600 (7)	--	

n = number of analyses.

retention time is long, the temperature difference between the content of
the aeration tank and the air is very small. This results in a big
difference in capacity between winter and summer. Under cold conditions not
more than about 35 m^3 per day is treated, whilst processing has to be
stopped when it freezes for more than a few days. The daily manure volume
to be processed increases up to 120 m^3 in the summer. The aeration capacity
is not limiting in this plant. In a year about 27 000 m^3 of manure can be
treated.

About 50% of the veal calves are raised in the area around Elspeet
(Veluwe). There are plans to install five more plants to treat about 85% of
the veal calf manure produced in this area (ca. 570 000 m^3/year). The
construction of the first has recently started; a second will follow this
year. The process will be discontinuous activated sludge as in the existing
Elspeet plant; the problem of low capacity in the winter will be (partly)
tackled by the installation of fine-bubble aeration; insulation of the
surface of the aeration tank will also be considered. This year the
Agricultural University of Wageningen will start with treatment by the
anaerobic upflow process followed by NH_3 stripping. Depending on the
results this process can be used in the last of the plants mentioned above.

4. PIG MANURE

Apart from a few purification plants for diluted pig manure at
individual farms, pig manure is not yet processed in the Netherlands. Since
1979 research institutes and universities have been working on processes
which are to be applied in large-scale plants, however, since it is
expected that, in the long run, several millions of pig manure will have to
be treated. Figure 4 and Figure 5 show two of the most promising schemes
currently under research. In the process of Figure 4 undigested manure or a
mixture of undigested and digested manure is conditioned with iron chloride
and polyelectrolyte and, after this, separated and dewatered by belt press
or decanter-centrifuge. The filtrate is purified in an activated sludge
process with emphasis on nitrogen removal. It is expected that, generally,
water authorities will strongly oppose the discharge of this effluent to
water courses. The answer is sought in an extra treatment step by reverse
osmosis. Dependent on the possibilities of the market for residual
products, the cake produced by belt press or centrifuge has to be dried or
incinerated, whilst the concentrate produced by the reverse osmosis step
may have to be concentrated or dried. Anaerobic digestion in this process
is only possible for a part of the manure because research has shown that

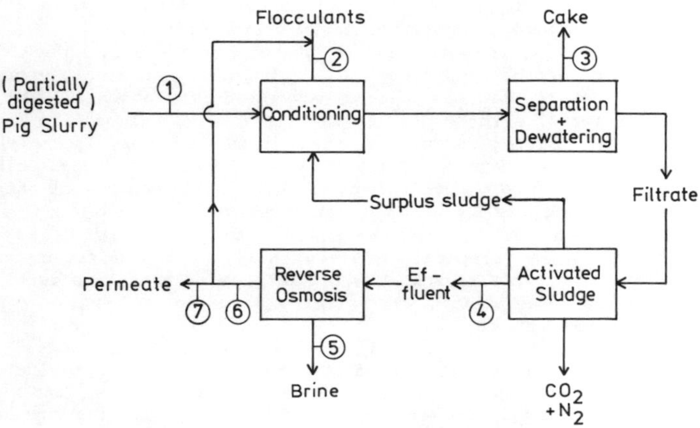

Figure 4. Process for the treatment of pig manure under research at the experimental pig station at Sterksel. The figures refer to Table 4.

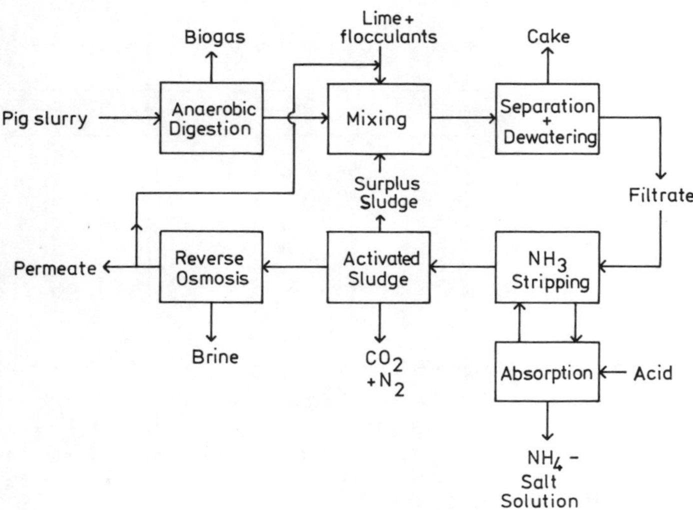

Figure 5. Process for the treatment of pig manure under research by the Agricultural University of Wageningen.

there is insufficient BOD left for a complete denitrification when all the manure is digested before the separation/dewatering step (2).

The other process, schematically pictured in Figure 5, starts with anaerobic digestion of the complete flow. By doing so, not only energy is generated, but also organic nitrogen is converted into ammonia. The next step is the addition of a rather large quantity (ca. 20 kg) of lime and the same or a somewhat smaller dosage of iron-chloride and polyelectrolyte as in the process described before. After this, separation and dewatering is carried out. The filtrate or centrifugate, high in pH because of the lime addition, is now stripped by large quantities of air, resulting in the transfer of the main part of the ammonia to the air. In an adjacent absorber the ammonia is trapped in a strong acid, e.g. sulfuric acid. A part of the remaining impurities in the filtrate or centrifugate is removed in an activated sludge plant. The final polish is a reverse osmosis treatment as described above.

The first process is studied by a farmer's co-operative, the Noord-Brabantse Christelijke Boerenbond NCB; the second by the Agricultural University of Wageningen. Various research institutes and other universities are involved, however. The NCB runs a pilot plant at an experimental farm at a village called Sterksel; this plant consists of an 80 m^3 digester, a 0.6 m wide belt press (Guinard) and an activated sludge plant with a 42 m^3 aeration tank. In a short time a reverse osmosis plant will be installed to evaluate and optimize bench scale research carried out by the TNO organisation.

Table 2 shows the results of experiments on the separation of undigested pig manure with the 'Sterksel' belt press. Conditioning was performed with the addition of 1 kg $FeCl_3$ and 0.25 kg Praestol 421K per m^3 of manure. Similar results were obtained with the separation of digested manure. The dosages of iron-chloride and polyelectrolyte could be reduced by about 30 and 50%, respectively, and Praestol 410K gave better results than the more cationic type 421K.

Table 2. Separation and dewatering of pig manure with a belt press at Sterksel.

Parameter		Slurry	Cake*	Filtrate*
Dry matter	%	9.7	27.0	1.9
Ash	"	3.1	7.7	1.1
Nkj	"	0.74	1.18	0.38
NH_4 -N	"	0.37	0.40	0.27
P	"	0.26	0.79	0.02
K	"	0.56	0.46	0.44
Fe	mg/kg	226	1 773	20.0
Cu	"	106	355	1.0

* After addition of 195 l water/m^3 slurry with 1 kg $FeCl_3$ and 0.25 kg Praestol 421K.

The optimisation of the biological treatment of the filtrate is now under research. One has to wait for detailed information until the results are published by the 'Sterksel' research group. However, some data regarding the effluent quality can be derived from a report of TNO about their experiments on reverse osmosis (3). Table 3 shows the results of the filtration of a sample of effluent from the 'Sterksel' plant through a composite membrane (ZF-99, PCI). Mostly the NH_4 concentration of the effluent is about zero, which means about 130 mg N-kj/l. By comparing the concentrations in the filtrate (Table 2) with those in the effluent of the biological stage (Table 3), it can be seen that considerable reductions in the values of several parameters are achieved. However, the absolute values of the concentrations are generally so high that it is expected that most water authorities will resist the discharge of large quantities of these fluids. Reverse osmosis improves the effluent quality considerably as can be seen in Table 3; the dark brown colour disappears almost completely.

Table 3. Composition of an effluent sample from the biological treatment stage of the experimental plant at Sterksel before and after reverse osmosis (ZF-99 composite membrane, PCI) and the composition of the brine.

Parameter		Effluent	Permeat	Brine
Volume ratio	--	100	75	25
COD	mg/l	2 285	15	9 120
BOD	"	31	<1	--
N kjeldahl	"	293	11	1 030
NH_4 -N	"	160	11	510
P	"	35	1.4	132
Cl	g/l	2.8 (1.8*)	0.12	10.5
SO_4	mg/l	184	12.5	--
Dry matter	g/l	11	0.6	42
Ash	"	7.7	0.3	34
Conductivity	mS/cm	12.5	0.75	45
pH	--	7 (8.9*)	6.5	7.6
Ca	mg/l	70	0.45	--
Na	"	570	--	1 770
K	g/l	4.6	0.23	16.2
Cu	mg/l	0.1	< 0.01	--

*Before acidifying to pH 7 with HCl.

The heart of the process pictured in Figure 5 is the stage where the ammonia is stripped from the filtrate and absorbed in acid. This step was studied, on a pilot scale, by the Agricultural University of Wageningen. Several types of packing were used. The results were published last year (4). The TNO organisation looked into the separation and the dewatering of the suspended solids; the results are rather similar with those of the 'Sterksel' project mentioned above (5).

5. <u>DISPOSAL OF RESIDUAL PRODUCTS</u>

When, in a few years time, about 570 000 m³ of veal calf manure will be purified in the Veluwe area, about 60 000 m³ of surplus sludge (5% dm) will have to be disposed of. It is not certain how difficult this will be. If necessary the volume can be reduced to about 20% with the help of dewatering devices. This problem is small compared to that of the disposal of the residual products of the processing of pig manure, however.

Table 4 gives a rough material balance of the 'Sterksel' process. If 1 m³ of pig manure (9.7% dm) is processed, about 0.3 m³ of belt press cake and 0.2 m³ of concentrate from the reverse osmosis treatment waits for further disposal; not more than about 0.5 m³ of effluent is produced. The cake contains most of the original dry matter, phosphate, calcium and copper. The main part of the soluble inorganic material is collected in the concentrate; it consists mainly of KCl. The 'Agricultural University' process also produced biogas and a concentrated solution of an ammonia

Table 4. Rough material balance of the process outlined in Figure 4.

Figure 4		1	2	3*	4	5	6	7**
Total mass	(kg)	1 000	200	310	890	220	670	470
Water	"	900	190	230	880	210	670	470
Dry matter	"	97	1.2	81	9.8	9.4	0.4	0.4
Ash	"	30	1	22	6.8	6.6	0.2	0.2
N kjeldahl	"	7.4	--	3.8	0.3	0.3	0.007	0.007
P	"	2.6	--	2.4	0.03	0.03	0.0009	0.0009
K	"	5.6	--	1.5	4.1	3.9	0.15	0.15
Fe	(g)	230	340	550	--	--	--	--
Cu	"	106	--	105	0.1	0.1	< 0.01	< 0.01

* Biological surplus sludge (ca 3 kg dm) excluded.

** Assumption: by re-use of permeat for dilution of polyelectrolyte the
material balance is not changed.

salt, whilst the added lime can be found in the cake. (In the 'Sterksel'
process most of the nitrogen is converted to N_2 gas and discharged to the
atmosphere.)

The ideas about the disposal of these products are roughly as
follows. Cake can be used as a fertilizer when organic material and/or
phosphate is wanted. Dehydration is maybe necessary when the distance of
transport exceeds several hundred kilometers. Another possibility is
incineration. What remains is about 20 kg of ashes. A controlled tip as a
final solution is not a very attractive option whereas the processing of
1 million of pig manure results in the production of 20 000 tons of these
ashes: space for these quantities will be hard to find and one has to take
measures to prevent leaching. As these ashes contain about 20% P_2O_5 there
is a theoretical possibility for usage in pig feed or in artificial
P-fertilizer instead of crude phosphates. Ammonia salts, derived from
organic manure should not be too difficult to sell; they can replace other
artificial N-fertilizers. Artificial K-fertilizers may be replaced by the
brine from reverse osmosis, probably after further concentration or
dehydration. Conclusion: by processing, the various minerals are separated
and concentrated; their disposal means mainly use as fertilizer in
situations where otherwise artificial fertilizers would be used.

Short term market research is necessary, both in the Netherlands and
abroad, to establish the validity of these ideas. The results of such
investigations will, without doubt, influence the flow sheets of the
processes discussed in these pages in such a way that residual products
that cannot be sold will not be produced.

6. COSTS

The costs for the treatment of veal calf manure amount to about
D.fl. 11 per m^3 for a plant with a capacity of about 100 000 m^3 per year.
This price includes D.fl. 3 per m^3 for the transportation to the plant. At
the gate of the plant in Elspeet the price is D.fl. 4 per m^3.

Only rough calculations exist for the costs of the processing of pig
manure. Calculations for a plant with a capacity of about 170 000 m^3 per
year and working according to the 'Sterksel' process, with incineration of
the cake and dehydration of the concentrate, resulted in a price of
D.fl. 25-35 per m^3, transport to and from the plant included.

These costs pose a very serious problem. It is not clear if they are
prohibitive. This will depend, among other factors, on the quantities

processed in relation to the possibilities to spread these costs over other farms and on the income from the sale of the residual products. For veal calf manure there is no alternative, however, due to the lack of demand. For pig manure, transportation from one side to the other side of the country is cheaper than processing (about D.fl. 20-25 per m^3). Treatment in a central plant is therefore expected to be practised only after all other possibilities are exhausted.

7. IN CONCLUSION

The Dutch research regarding the purification/treatment of low concentrated animal slurries dates back to 1965. At first only one stage treatment with very low loaded activated sludge systems was studied and applied. In the seventies all attention was focussed on the optimization of nitrogen removal. The results were used in the design of the central purification plant for veal calf manure at Elspeet which was commissioned in 1976.

After the first experience with the one step biological purification of manure it became clear that it is not effective to leave out pretreatment if there is more than 1-2% dm in the slurry. During the last few years attention has been focussed for these reasons on separation and dewatering. Apart from this the opinion was formed that in many situations, especially when large volumes are to be treated, biological purification of manurial fluids will not be enough. Bench scale experiments with reverse osmosis showed promising results, but this unit operation is not yet available for use on a full-scale.

Large-scale processing of pig manure will really be possible only when there is no doubt about the disposal of the residual products, which means where, in what form and for what price. This aspect deserves full focus in order to prevent a major incident.

REFERENCES

(1) WIJNANDS, J.H.M. and LUESINK, H.H. (1984). Een economische analyse van transport en verwerking van mestoverschotten in Nederland. Research report nr 12, Landbouw-Economisch Instituut, Den Haag.

(2) TEN HAVE, P.J.W. (1983). Aerobe zuivering van varkensdrijfmest na anaerobe behandeling. H_2O nr 6, pp 120-124.

(3) VAN TONGEREN, W.G.J.M. and VAN VEEN, H.J. (1984). Onderzoek naar de toepassingsmogelijkheden van omgekeerde osmose bij de zuivering van varkensdrijfmest. Report nr 83-015272, Hoofdgroep Maatschappelijke Technologie-TNO, Apeldoorn.

(4) SCHOMAKER, A.H.H.M., BEVERLOO, W.A. and VAN VELSEN, A.F.M. (1984). Procestechnologisch onderzoek van het NH_3 luchtstrip/absorptieproces. Report Landbouwhogeschool Wageningen, vakgroep waterzuivering.

(5) VAN VEEN, H.J. (1983). Ontwatering van varkensdrijfmest. Report nr 83-04380, Hoofdgroep Maatschappelijke Technologie-TNO, Apeldoorn.

DISCUSSION

A. Wellinger
Did the calf manure treated contain straw?

Wouldn't the process be less expensive if the manure was diluted with straw?

P.J.W. ten Have
No, calves are kept in wooden boxes without straw.

I don't think so. It could affect the whole economics.

L.P. Savelkoul
Would it be possible to treat domestic waste together with the animal waste? I do realise though that it would depend on the final destination of the sludge produced.

It could be one option. However, it is important to consider every option. These can't always be considered in the abstract. You have to carry out an option to really find out the true costs. This is especially the case as the easy option of disposal to agricultural land is being restricted.

T.J. Casey
It seems to me that the phosphorus content of the waste is the real limiting step to disposal. Certainly, aerobic processes can be used for removing nitrates whilst this is not really feasible with anaerobic processes. Even if you were to use reverse osmosis, there would be problems with membrane fouling.

Yes, phosphorus, which can lead to eutrophication, is the major problem. Nitrates are also a problem especially if land disposal is con-sidered. Reverse osmosis has been tried and there weren't as many problems encountered as might have been expected. With anaerobic processes you still have to add an extra carbon source, for example, methanol. Ammonia stripping has also been tried.

P. Balmer
Is the problem with eutrophication a very severe problem?

Yes, it can be.

A.M. Bruce
Is there an export market for the waste?

I would have thought that you could have real odour problems. Further-more, I would be concerned that the ammonia loss suggested that you could also have problems with acid rain.

That is one further option that we have considered.

Yes, these could be real problems, especially the odours.

H. Scheltinga
The problems described by Mr. ten Have are very serious at the present time in the Netherlands. It also has to be remembered that any reduction in the animal population in the country would have enormous social effects.

Yes, the problems are very serious and the trends suggest that they could get worse.

<u>SESSION II</u>

CHAIRMAN: Mr. A.M. BRUCE

Rapporteur: Dr. P.J. NEWMAN

AEROBIC THERMOPHILIC DIGESTION OF PRE-THICKENED SLUDGE USING AIR

B. PAULSRUD
Aquateam - Norwegian Water Technology Centre A/S
G. LANGELAND
Norwegian College of Veterinary Medicine

Summary

Pilot scale studies of (the JANCA-process) aerobic, thermophilic digestion of pre-thickened sludge using air were performed for seven weeks at HIAS sewage treatment plant in Norway. By batch operation of the process with sludge draw and fill every second day, temperatures above 60°C could be maintained in the single stage reactor for approximately 30 hours. The treated sludge was satisfactorily hygienized, but the detention time was too short for rendering the sludge stable and odour free.

The thickening device of the process (a vibrating sieve) did not function properly with the mixed primary-activated-chemical (Al) sludge and should therefore be replaced by another type of simple equipment that can produce a sludge of 8-10% dry solids (e.g. dewatering container).

The operating cost of the process (energy and polymer) is estimated to be about 150 NOK/tonne DS, and the capital cost for a plant of 2 100 tonne DS/year is about 2 mill. NOK (1984).

1. INTRODUCTION

Several methods for hygienization of sewage sludge have been evaluated in Norway: lime treatment (1), composting (1), long-term storage (2) and aerobic, thermophilic digestion using pure oxygen (3,4). In 1983 the Danish company JANCA introduced on the Norwegian market a new version of the aerobic, thermophilic digestion process using a special device (vibrating sieve) for thickening the raw sludge up to 8-10% DS before digestion. With this concentrated sludge, a closed and insulated reactor and oxygen supply from a surface aerator, it was claimed that the sludge would be hygienized with a retention time of about one day (batch operation). This was based on full-scale experiences from Denmark with the system operating as the aerobic stage of the 'dual digestion' process.

Previous work with aerobic, thermophilic digestion in Norway has been devoted to a pure oxygen process (3,4). Consequently, there was great interest in getting more data about the JANCA-process to make a comparison of both performance and cost. A pilot plant study of the JANCA-process was set up at the HIAS sewage treatment plant; the same place where the full-scale demonstration study of the pure oxygen process was performed during 1983 (4).

2. MATERIALS AND METHODS

2.1 Pilot Plant Description

The pilot plant consisted of a thickening device (vibrating sieve) with polymer dosing equipment and a single stage digester with a surface aerator (see Figures 1 and 2).

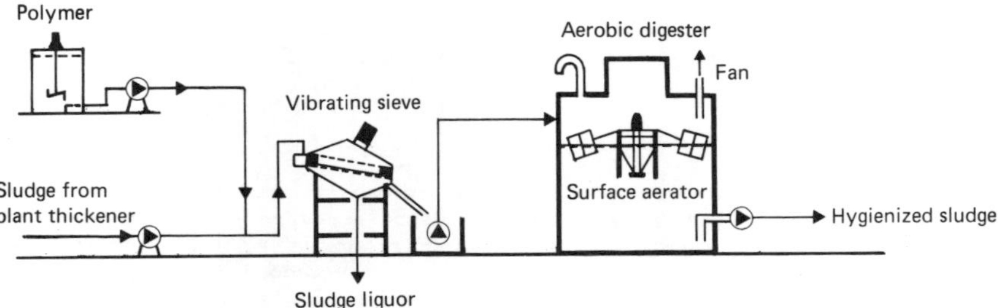

Figure 1. Flow diagram of the JANCA pilot plant.

Figure 2. The pilot plant with the polymer tank in front, the digester at
the back and the vibrating sieve in between.

Mixed primary-activated-chemical (Al) sludge was pumped from one of the treatment plant thickeners to the vibrating sieve. Polymer (0.1% solution) was added to the sludge in a mixing chamber and the conditioned sludge was spread onto the inclined sieve, which was vibrating with an adjustable frequency. A solid-liquid separation took place while the sludge moved down the sieve. The solid phase was collected in a small tank and then pumped to the digester, while the sludge liquor was transported back to the treatment plant inlet. The capacity of the pilot scale sieve was about 1 m^3/h.

The digester was a cylindrical steel tank insulated with 10 cm mineral wool and with a manhole on the top. Maximum effective volume of the tank was 4 m^3. The 3 kW surface aerator was floating on three pontoons, and start/stop of the aerator could be controlled by the reactor temperature (not used during the study). A fan was placed in the exhaust pipe to increase the air flow through the reactor.

2.2 Pilot Plant Operation

The digester was fed on a draw and fill basis (batch operation) with the time between each feeding of sludge as the main variable during the study. Normally about 40% of the sludge volume was left behind in the reactor when pumping out the treated sludge.

The study was divided into three phases (see Table 1).

Table 1. Phases, duration and operational conditions

Phase	Duration	Operational conditions
Start-up	24 April–12 May 1984	Sludge withdrawal from the digester <u>every</u> morning. Sludge feeding in variable amounts (0.4-2 m^3) during 1-4 hours.
Test period 1	14-29 May 1984	Sludge withdrawal from the digester <u>every</u> morning. Sludge feeding (approx. 2 m^3) during 1-3.5 hours.
Test period 2	1-19 June 1984	Sludge withdrawal from the digester <u>every second</u> morning. Sludge feeding (approx. 2 m^3) during 2-3 hours.

2.3 Monitoring, Sampling and Analyses

The flow rates of sludge, sludge liquor and polymer were monitored, and so were the temperature in the digester and the total power consumption for the aerator, vibrating sieve and sludge pumps.

Samples were taken manually every time sludge was pumped to the vibrating sieve and when entering and leaving the digester. Several grab samples were mixed to make one composite sample. Some days the sludge liquor from the sieve was also sampled in the same way.

All the sludge samples were analysed for total solids (dry solids) and volatile solids, while suspended solids and BOD_7 were used to characterize the sludge liquor from the vibrating sieve.

Once a week grab samples of the sludge entering and leaving the digester were taken for microbiological examinations. These were performed by the Norwegian College of Veterinary Medicine and included faecal coliforms, faeca streptococci, spores of <u>Clostridium perfringens</u>, <u>Salmonella</u> bacteria, and bacteriophages (added to the raw sludge). The microbiological methods are described in the work by Langeland et al (4).

3. RESULTS AND DISCUSSION

3.1 Pre-thickening

During the start-up period several polymer types and sieves with different apertures were examined in order to increase the dry solids content of the pre-thickened sludge. Despite all efforts dry solids of 8-10% could only be attained for a few days with the sludge from the HIAS plant. Figure 3 shows the variation in dry solids content of the sludge before and after the thickening device.

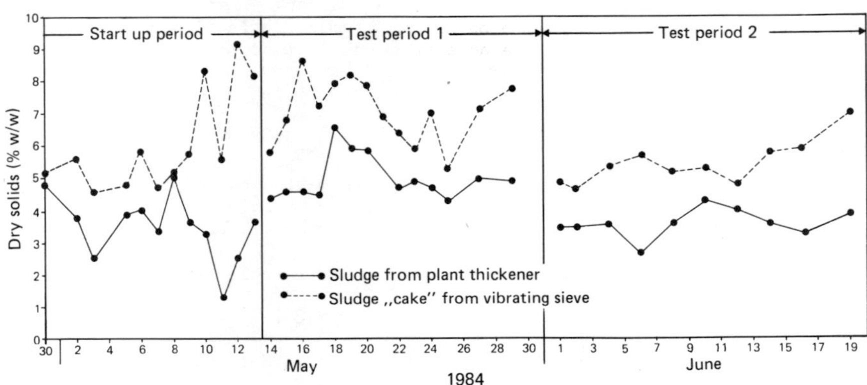

Figure 3. Dry solids content of the sludge before and after pre-thickening with the vibrating sieve.

In Table 2 there are summarized some data (mean values) from the operation of the vibrating sieve.

Table 2. Mean values for dry solids concentration of sludge, polymer dosage and quality of sludge liquor

Phase	Sludge from plant thickener (% DS)	Sludge after sieving (% DS)	Polymer dosage (kg/tonne DS)	Sludge liquor* Suspended solids (mg/l)	BOD_7 (mg/l)
Start-up	3.5	6.1	4.8	2 000	2 000
Test period 1	5.0	7.1	3.7	4 000	1 000
Test period 2	3.6	5.5	3.6	4 500	2 700

* The data for sludge liquor quality are based on very few samples and the values should be considered accordingly.

In addition to poor thickening efficiency and poor sludge liquor quality, the vibrating sieve also required close attention by an operator to function properly. All these factors made it easy to conclude that this equipment was unsuitable for pre-thickening the sludge at HIAS treatment plant.

Another type of pre-thickening device was tested on the HIAS sludge some months after the pilot plant study. This was a gravity type dewatering container (with a special filter cloth on the walls) which has proved to be well suited for dewatering polymer-conditioned septage (5). With the mixed primary-activated-chemical sludge at HIAS there seemed to be no problem in concentrating the sludge from the plant thickeners up to about 10% dry solids with a polymer dosage of approximately 3 kg/tonne DS.

3.2 Loadings and Temperatures

Due to the variations in solids concentration of the pre-thickened sludge (Figure 3) and the operational problem with the thickening device, both organic and volumetric loadings on the digester varied considerably during the study. Table 3 gives data on retention time and organic loading for the test periods.

Table 3. Loading Data for the Digester

Phase	Retention time* (days)		Organic loading (kg VS/m^3/d)
	Range	Mean	Mean
Start-up	2.1-3.6	2.8	16
Test period 1	1.5-2.2	1.9	27
Test period 2	2.4-4.0	3.1	11

* Retention time is calculated as the sludge volume in digester after feeding (m^3) divided by the sludge feed rate (m^3/d).

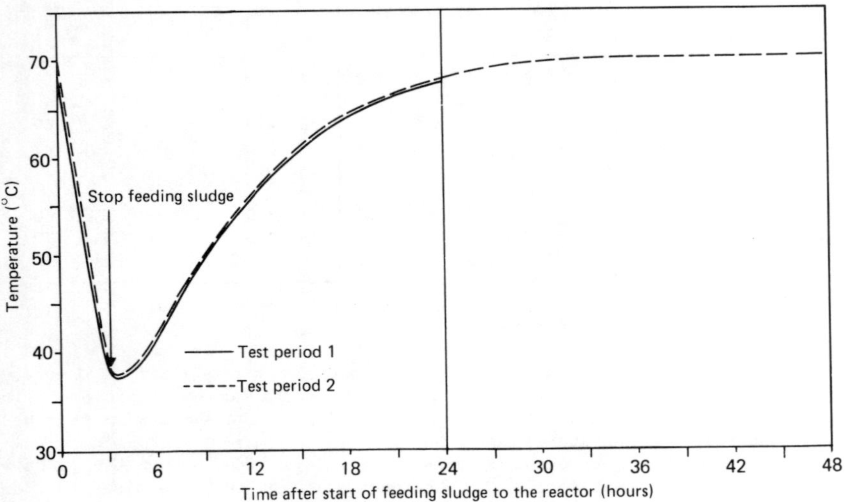

Figure 4. Typical temperature curves in digester after start of feeding sludge.

Sludge temperature in the digester could hardly be raised above 58°C in the first part of the start-up period. This problem, however, was solved when the air flow rate through the reactor was reduced from 100 m³/h to 20 m³/h by cutting out the fan and reducing the diameter of the air intake. After this correction the digester temperature normally followed the curves shown in Figure 4.

In test period 1 the process could maintain temperatures above 60°C for 8-12 hours until the next sludge feeding decreased the temperature to 35-40°C. By feeding the digester every second day (test period 2), the sludge temperature was above 60°C for 27-34 hours and in about 20 of these hours the temperature was approximately 70°C.

These results were obtained with a feed sludge of lower solids concentration than anticipated. It is still a question whether a feed sludge of 8-10% dry solids might involve a more rapid increase in digester temperature after the end of sludge feeding and thereby increase the time of high temperature exposure for a given retention time.

3.3. Hygienization

The results of the microbiological examinations are given in Table 4.

Table 4. Concentration of test bacteria and added bacteriophages in feed sludge and digested sludge. (All concentrations given as the logarithmic value of the number of organisms per gram dry solids. FC: faecal coliforms, FS: faecal streptococci, SCP: spores of Clostridium perfringens, SALM: Salmonella bacteria, BP: bacteriophages (coliphage MS 2), TEMP: sludge temperature when sampling, -:not detected.

Phase	Sampling date 1984	Feed sludge					Digested sludge					
		FC	FS	SCP	SALM	BP	FC	FS	SCP	SALM	BP	TEMP
Start-up	05.07	5.4	5.9	5.4	2.7	4.2	3.9	4.9	6.4	–	4.3	57
Test period 1	05.14	6.0	7.1	6.4	3.3	4.5	3.5	4.4	5.5	–	2.7	69
	05.21	5.8	7.1	6.4	3.4	4.9	6.0	7.0	6.1	3.2	5.4	67
	05.28	6.3	7.1	6.5	3.9	4.8	2.6	4.4	6.2	–	3.1	69
Test period 2	06.06*	6.7	6.8	6.3	3.6	4.8	2.7	3.9	4.0	–	2.7	60
	06.12	7.3	6.7	6.5	3.3	3.1	–	2.3	4.3	–	2.0	70
	06.18	5.8	6.8	6.5	3.6	3.9	–	–	4.0	–	–	70

* This was not representative for test period 2 due to aerator malfunction prior to sampling.

The operating conditions in test period 1 (draw and fill every morning, average retention time: 1.9 days) gave a limited improvement in sludge hygienic quality. This was probably due to the short time exposure (8-12 hours) of temperatures above 60°C.

With an average retention time of 3.1 days and feeding the reactor every second day (test period 2), the decimation of test organisms were so good that, comparing with Norwegian guidelines, the sludge could be spread on land without any further treatment or storage.

3.4 Stabilization

Due to short test periods and considerable day to day variations in solids content of feed sludge, reliable values for organic matter reduction during digestion were difficult to achieve. Calculations based on the approximate mass balance method (6) indicated a volatile solids destruction of 20-25% for both test periods. This is in close agreement with results obtained in the pure oxygen project with three days retention time (4).

With this organic matter reduction, you should not expect a stable and odour free sludge from the reactor. This was confirmed in practice by a very unpleasant smell from anaerobic storage of the sludge prior to dewatering.

3.5 Power Consumption

The total power consumption for aeration, pre-thickening and sludge pumping was fairly constant at 78 kWh/d throughout the period of investigation. This means a specific power consumption of about 30 kWh/m^3 sludge from plant thickener in test period 1, and about 42 kWh/m^3 in test period 2.

These figures are higher than expected for a full-scale plant where the use of automatic temperature control can reduce the running hours of the aerator to the necessary minimum for maintaining reactor temperatures of, for instance, 65°C. Furthermore, the size of the surface aerators in a full-scale plant can be better adjusted to the volume of the reactor.

Reported power consumption for aerobic, thermophilic digestion with air, is in the range of 12.5-19 kWh/m^3 (7, 8, 9), and it is claimed by the manufacturer that a full scale JANCA plant can compete with these figures.

4. COST

The cost figures given below are based upon the pilot plant study and the bids given by the JANCA company for a full-scale "hygienization" plant (2 100 tonnes DS/year) at the HIAS sewage treatment plant. The quoted plant consisted of containers for pre-thickening of sludge and two 80 m^3 reactors, each with a surface aerator of 30 kW. The operating mode should be the same as in test period 2, i.e. draw and fill of each reactor every second day.

Capital cost: NOK 2 millions.

Operating costs (energy and polymer): NOK 150/tonne DS.

5. CONCLUSIONS

* The JANCA-process with sludge draw and fill <u>every</u> morning (average retention time 1.9 days) could raise the sludge temperature to above 60°C and maintain this level for 8-12 hours. This caused a limited improvement in the sludge hygienic quality.
* With sludge draw and fill <u>every</u> <u>second</u> morning (average retention time 3.1 days) temperatures above 60°C could be maintained for 27-34 hours, and the sludge was satisfactorily hygienized.
* The data on organic matter reduction during the process are uncertain, but figures of 20-25% are indicated. This means the treated sludge is not stable and can create odour problems during subsequent handling and disposal.
* The original pre-thickening unit (vibrating sieve) did not function properly with the mixed primary-activated-chemical sludge. Another device (dewatering container) seemed to be capable of concentrating the sludge up to about 10% dry solids.
* Total power consumption for the pilot plant was 30 and 42 kWh/m^3 sludge from plant thickener with sludge feeding every day and every second day, respectively.
* Capital cost for a 2 100 tonnes DS/year plant is about NOK 2 million (1984). The operating costs (energy and polymer) will be about NOK 150/tonne DS.

REFERENCES

(1) PAULSRUD, B. and EIKUM, A.S. (1984). Experiences with lime stabilisation and composting of sewage sludge. In Sewage Sludge Disinfection and Stabilisation (Ed. A.M. Bruce). Ellis Horwood Publishers Ltd., Chichester, England, 261-277.

(2) LANGELAND, G. (1981). Noen hygieniske forhold ved kloakkslam. Tidsskrift for Den norske laegeforening, 101, 1276-1278.

(3) PAULSRUD, B., HAUGAN, B.-E, and LANGELAND, G. (1984). Aerobic, thermophilic digestion of sewage sludge using pure oxygen. In Processing and Use of Sewage Sludge. (Ed. L'Hermite, P. and Ott, H.). Commission of the European Communities, D., Reidel Publishing Company, 107-109.

(4) LANGELAND, G., PAULSRUD, B. and HAUGAN, B.-E., (1984). Aerobic, thermophilic stabilization. Paper presented to COST 68-seminar. Inactivation of Microorganisms in Sewage Sludge by Stabilization Processes, Hohenheim/Stuttgart.

(5) PAULSRUD, B. (1982). Avvanning av septikslam i container. Report no. 4/82, Norwegian Institute for Water Research, Oslo.

(6) FISCHER, W.J. (1984). Calculation of volatile solids destruction during sludge digestion. In Sewage Sludge Disinfection and Stabilisation (Ed. A.M. Bruce). Ellis Horwood Publishers Ltd., Chichester, England.

(7) MORGAN, S.F., GUNSON, H.G., LITTLEWOOD, M.H. and WINSTANLEY, R. (1984). Aerobic thermophilic digestion of sludge using air. In Sewage Sludge Disinfection and Stabilisation (Ed. A.M. Bruce). Ellis Horwood Publishers Ltd., Chichester, England.

(8) JEWELL, W.J., KABRICK, R.M. and SPADA, J.A. (1982). Autoheated, aerobic thermophilic digestion with air aeration, EPA Project Summary, EPA-600/52-82-023.

(9) LOLL, U. (1981). Betriebswerte und wirtschaftlichkeit der Aerob Thermophilen Stabilisation. In Biotechnologie im Abwasserbereich, BMFT-Statusseminar.

DISCUSSION

U. Loll
You mentioned an average retention time of 3.1 days. I would have thought a minimum of 5 days would be better especially if the temperature is not high enough. You could try pre-thickening, say up to 10-12% total solids, but then, there could be problems with aeration. If thicker sludges are used you need a longer retention time because of course, the organic content is also increased. On balance, I don't think, though, you should go above 8% total solids.

B. Paulsrud
Thank you for these comments. I note your point about thick sludges.

P. Balmer
Do you think that there will be serious odour problems with the

We didn't estimate the level of the odour problem but I imagine that it

process? What were the qualities of the sludge liquors? Were any bacteria or spores added to the sludge?

could be quite bad. The quality of the sludge was quite poor and although the sampling was inadequate, values as high as 4-5 000 mg/l suspended solids were observed. No bacteria or spores were added to the sludge. We did add a virus indicator though.

A. Wellinger
With cattle manure at 8-10% solids, I have found that an immersed aerator is more efficient than a surface aerator.

We haven't tried immersed aeration.

T.J. Casey
How long did the draw and fill process take? Were the aerators working during this time? If so this could have a large effect on the bacteria present and, of course, on the temperature.

The draw and fill process takes about 3-4 hours. The aerators were working continuously and we had no means of temperature control.

U. Loll
There are about 20 plants operating in the Federal Republic which use the aerobic, thermophilic process and with adequate insulation can easily reach 60-70°C. If you add the new batch of sludge in about 30 minutes the temperature in the digester may fall to 30°C but it rises very quickly back to 60-70°C as the bacteria are very accommodative. If the process is run correctly, there needn't be too many odour problems. However, if the pH rises up to 8.9, then there can be problems with ammonia release.

I think the point about the speed of draw and fill is very important and, in some part, explains why our digester tended to run at a lower average temperature.

A.M. Bruce
Is the process used in the Netherlands? It is being used in the UK for sewage sludges and pig slurry.

H. Scheltinga
I don't think it has been used in the Netherlands.

I am surprised that the process doesn't kill the salmonella especially since the temperature achieved and the retention time is long enough.

B. Paulsrud
Yes, it should be effective. Perhaps our problems were a result of by-passing or thick sludge present. Of course, the lower operating temperature could have been the cause.

P. Ockier
What is the optimum size for aerobic, thermophilic plant?

A.M. Bruce
About 5-20 000 population equivalents.

U. Loll
I think that the upper figure could be up to 30 000 or even 50 000 population equivalents.

P. Balmer
A number of retention times for effective stabilisation have been mentioned. I believe that you should use at least 6 days with aerobic digestion and about 15 days for anaerobic digestion. Of course, the temperatures reached in the digesters would affect the times considerably.

Thank you for your comment. Certainly the average retention time we used was considerably less than you recommend.

PROCESSING ORGANIC WASTE PRODUCTS TO BLACK SOIL AND ORGANIC FERTILIZERS

Ir. J.G. ten WOLDE
Rutte Recycling B.V.,
AMSTERDAM - The Netherlands

Summary

Processing organic waste products into black soil and organic fertilizers, apart from the possibilities of incineration and dumping, is one of the alternatives for sludge processing. Measures to promote recycling are stimulated by the government. For economic reasons, recycling merits maximum attention. On the basis of the fertilizer characteristics of purification sludge and other organic waste products (agro-industrial by-products) processing into fertilizer and soil are regarded as possibilities, provided the final product meets certain quality criteria. In the Netherlands standards are being developed for the basic soil quality and for organic fertilizers. By combining "loaded" and "unloaded" organic waste products specified qualities of organic fertilizers and black soil are attainable. Black soil is obtained by mixing sandy soils with stabilized organic materials in order to obtain better structural properties for improved growth conditions for landscaping. Stabilization of purified sludge takes place on natural drying beds and by composting in covered halls. The composition to be realised is determined on the basis of available analytical data by means of a computer projection. The black soil is applied in various green belt projects. The organic fertilizers are used in agriculture.

1. INTRODUCTION

In a recent number of The Journal of Waste Recycling my attention was attracted by an article entitled "Somersaults in sludge policy", contents of which boiled down to the following. In 1976 enormous quantities of sludge were found on the beaches of New York; the sludge originated from the regional water purification plants as a result of sea disposals through a pipeline. The local population rebelled and the federal EPA (Environmental Protection Agency) reacted almost immediately by ordering the regional managers of the purification plants at short notice to start making a plan for sludge treatment with an agricultural destination. Two managers of the controlling organization reacted as follows: one, "that it has never been proved that sea disposal before the city of New York involves any negative consequences"; the other, "it has never been proved that an agricultural destination is a definite solution of the sludge problem". Both remarks are correct in my opinion. I could add a third: "It has never been proved that waste incineration is a solution without environmental problems (final destination of ash, air pollution)". In fact, the example makes it clear that the sludge producers are at present in a difficult position. In the field of environmental hygiene we find ourselves in a continuous stream of findings and meanwhile the sludge producer is up to his ears in sludge. However, I would first draw one conclusion from the above remarks: uncontrolled dumping, wherever it may be, is absolutely a thing of the past; but the present scientific agricultural know-how must enable us to use waste products correctly as fertilizer.

2. COMPARING BLACK SOIL AND FERTILIZER PREPARATION VERSUS ALTERNATIVE WASTE PROCESSING METHODS

Waste occurs in three physical forms: solid, liquid and gases. It may in general be assumed that solid waste is fairly easy to control. In this physical state storage is simpler than in the liquid state, storage in the gaseous phase being relatively complicated. When solid waste is incinerated, pollutants in the gaseous phase will thus find their way into a relatively large volume. The air purification problem is therefore complicated. I do not want to say that in future we should give up incineration, but that we must all try preferably to produce waste that can be recycled in the solid phase. In fact, in the past few years the purifying centres have also tackled the water pollution problem by converting the liquid waste components into the solid phase. This line of thought is now also generally followed by the EPA as their policy: "Try to solve the waste problem via the solid phase".

In this physical state the products are easy to control, which is also the starting point in the policy of the Ministry of Public Health and Environmental Hygiene. A similar approach, in order of priority, is encountered for domestic waste processing:
1.   Sanitation by the producer;
2.   Measures to promote recycling;
3.   Incineration of any part that remains;
4.   Dumping on dumping grounds.

Irrespective of the form in which waste is produced, there are two main solutions for waste processing: recycling and destruction, starting from the assumption that dumping has a low priority. When making a choice between waste processing systems attention should be paid to the factors, capital, labour and market expectations for the positive waste components (such as fertilizers and energy).

Table 1 gives an approximate macro-comparison of the two solutions. Capital expenditure, labour and auxiliary materials for recycling are given very pessimistically, while for the sake of a not too optimistic comparison in favour of recycling, the estimated relative cost factor for destruction is low. In determining the yield factors we started from the average nitrogen and phosphate contents in purified sludge and the present market price (D.fl. 1.20 and D.fl. 1.40 per kg, respectively).

The energy production by incineration is based on a combustion value of 700 kW per ton of dry material at a yield of D.fl. 0.12 per kW. This combustion value is only feasible if the starting product completely consists of dry material; hence when incineration is applied, very close attention should be paid to dehydration or drying of the waste products. Any rise in the price of energy would result even more strongly in an advantage for the fertilizers, because the cost price of nitrogen is more than proportionally dependent on the rise in price of energy.

It is a very approximate comparison; no account has been taken, for example, of other environmental requirements. The only fact I want to stress is that there is ample room for financial manoeuvring for optimisation of the recycling system. When capitalized for the country's sludge products (5 000 000 m^3 is 4%) the annual operating difference is about 60 million guilders in favour of the recycling system. Another important factor for sludge processing is the way the sludge is dehydrated.

Table 2 shows the cost of various sludge dehydration systems and liquid agricultural offtake.

These data have been derived from a study of the installation for 80 000 pollution equivalents. The bottom figures in the columns can also be regarded as cost (in guilders) per ton of dry material, which clearly shows that natural dehydration opens up favourable perspectives to create

Table 1

Comparison of organic waste treatment systems

	Organic waste	
Agricultural reutilization		**Combustion**
Capital- labour-aid matter costs	− Dfl 300*	− Dfl 500*
Value of fertilizers or energy	+ Dfl 177*	+ Dfl 84*
Net costs	− Dfl 123*	− Dfl 416*

* Costs/ton dry matter

Table 2

Cost of sludge dewatering systems

x Dfl 1000	Fluid agri-cultural utilization	Dewater-ing by lagoons	Belt filter presses	Plate filter presses
Investment	360	1.150	1.510	2.390
Capital costs	13	39	97	159
Operation costs	195	70	327	225
Total year costs	208	109	424	384

Table 3

Summary of standards of micro-elements in sludge, organic fertilizers and soils (ppm in dry matter)

Micro element	Dutch sewage sludge (1982)	Soils		Organic fertilizers	
		Dutch A-value	Germany (Prof. Kloke)	Short* term	Long* term
Zinc	1650	200	300	900	250
Copper	492	50	100	300	50
Lead	385	50	100	750	150
Nickel	60	50	50	50	50
Chromium	220	100	100	150	100
Cadmium	8	1	3	3	1,5
Mercury	2,2	0,5	2	3	1,5
Arsenic	−	20	−	20	20

* Non-official

financial potentialities for destination optimalization of organic sludge products.

3. IS RECYCLING ECONOMICALLY VIABLE?

A recent publication of the NVA sludge inquiry shows that in previous years most sludge has found a positive (i.e. soil destination) application, the black soil preparation having taken a large share for its account. Properly speaking it is not surprising that landscaping attracts a fair amount of attention: the above survey shows that with respect to cost and on the basis of future expectations, agricultural use clearly offers better perspectives than destruction by incineration. To arrive at economically viable recycling in agriculture, the Union of Polder Boards has drawn up guidelines which define quantitative and qualitative standards for the composition and application of sludge.

The Ministry of Public Health and Environmental Hygiene, considering the approximate composition of biological sludge from sewage purification plants, claims that sludge has valuable properties (organic material, phosphate and nitrogen) as soil-improving material and fertilizer. Recycling of this sludge therefore fits in the general policy of the Ministry concerning waste products. A drawback and possibly even a factor prohibiting recycling of sludge is, for example, the presence of pollutants, notably heavy metals and persistent, toxic compounds.

In other words, all the more reason for sanitation at the source. Various agricultural research institutes in the EC Member States have paid very great attention to this problem. In addition to a directed agricultural policy, sanitation at the source will have to make correct agricultural application in future options. Seeing that a number of sludge grades failed to meet the Dutch standards and the fact that increasingly large quantities of sludge were produced which could not be handled quantitatively in agriculture, alternatives had to be sought at short notice. This necessity became even stronger as increasing demands were made on dumping grounds with respect to the consistency of the material to be dumped. The black soil industry proved to be one of the alternatives. It was supposed to give rise to fewer risks, because the black soil would not be directly used for agricultural production.

However, also on the basis of soil protection, reality dictates that the formulated soils should also obtain an agricultural destination in future. On behalf of soil protection the provisional Indicative Multi-Year programme Soil 1984-1988 incorporates soil criteria. This programme is going to be the basis for the Soil Protection Act and the Fertilizer Act. The Multi-Year programme includes a so-called A-value, which can be regarded as a natural background concentration of various elements and compounds. Black soil should meet this A-value. On the basis of this quality criterion black soil is in principle suitable for multi-functional use. This idea of basic standards for soil has been developed by Prof. Kloke (West Germany). Against this background of quality criteria the present black soil preparation may be regarded as justified, provided sludges and other organic waste products (such as agro-industrial by-products) are used which also meet given quality criteria.

A similar release policy is now being developed by the Ministry of Public Health. Additional scientific research could, for some elements and/or compounds, open the way to widening or narrowing the standards somewhat. Table 3 lists the present guidelines (not yet official) for black soil and organic fertilizers. The Soil Protection Act and the Fertilizers Act will have to define the future manoeuvring room (Figure 1). By combining "loaded" and "unloaded" organic (waste) products criteria for the quality of organic fertilizers and black soil can be realized.

Figure 1

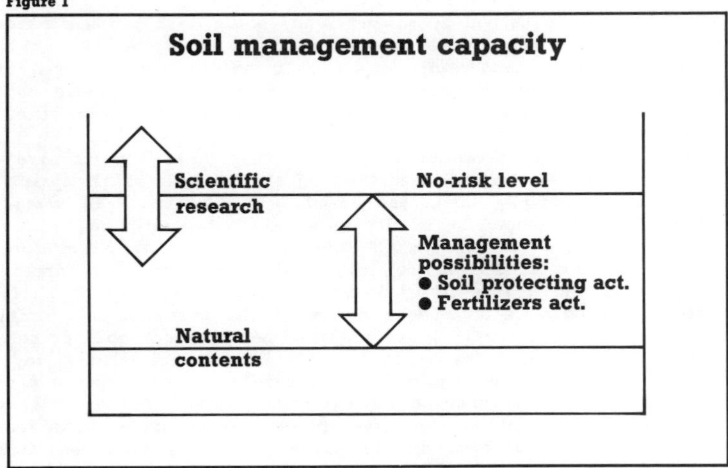
Soil management capacity
Scientific research
No-risk level
Management possibilities:
● Soil protecting act.
● Fertilizers act.
Natural contents

Figure 2

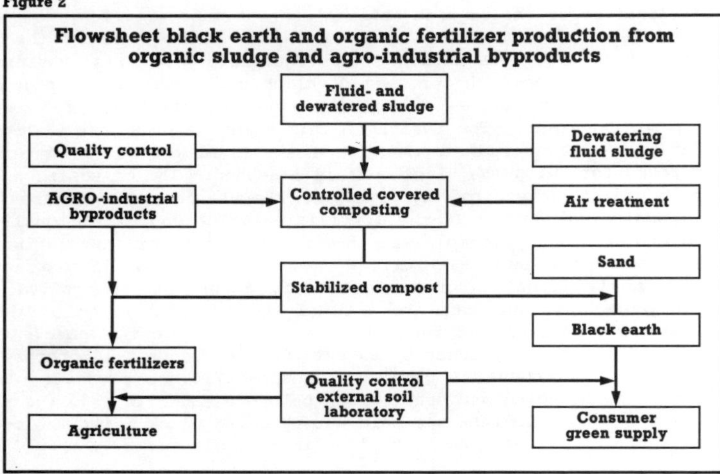
Flowsheet black earth and organic fertilizer production from organic sludge and agro-industrial byproducts
Fluid- and dewatered sludge
Quality control
Dewatering fluid sludge
AGRO-industrial byproducts
Controlled covered composting
Air treatment
Sand
Stabilized compost
Black earth
Organic fertilizers
Quality control external soil laboratory
Agriculture
Consumer green supply

4. PREPARATION OF BLACK SOIL AND ORGANIC FERTILIZERS

Black soil is obtained by mixing sandy soils with organic materials thus to improve the structural properties in order to arrive at better growth conditions for landscaping, etc. To ensure the required oxygen ratios in the black soil a certain maximum quantity of organic material should be adhered to (ca 8%), starting from thoroughly post-stabilized sludge.

Figure 2 shows the production scheme. The processes to prepare black soil differ in the dehydration condition of the sludges at the point when mineral additives (mostly sand) are added. In practice this means that either liquid sludge (about 5% of dry material) or pasty sludge (at least 25% of dry material) is mixed with sand. The starting point for both systems is that the sludge is not mixed with sand until a certain dry material content and degree of stabilization have been reached. If the dehydrated or incoming sludge does not meet the stabilization criteria, additional treatments (aeration) are required before black soil preparation proper can take place. The initial condition of the sludge may vary appreciably. In the case of liquid sludge, dehydration takes place first of all to ca. 25% of dry material on natural drying beds. During this drying process initial stabilization also takes place. These drying beds are then cleared and the dehydrated sludge is placed on piles in order to allow post-stabilization. Dependent on the quality of the sludge and the requirements the final product has to meet, other organic material can be added when the drying beds are cleared. Even if the incoming sludge is already sufficiently dehydrated, the material may still have to undergo a treatment to reach the required stabilization. The treatment selected is highly dependent on the history of the sludge.

Very recently our production unit for this purpose started an extensive composting process according to the aerated-pile method. Using mechanical turning and mixing machines, continued composting and mixing and thus stabilization of the organic material is obtained. Besides, composting affects pasteurization of the sludge. Once the sludge, after the above pre-treatments if any, has attained the required dry material content and degree of stabilization, the process of black soil preparation proper can take place. For this purpose sludge and sand are mixed in the required ratio in a mixing plant. The design of this plant is such as to allow any mixing ratio. The required composition of the final product is determined by the purpose for which the black soil is intended. The composition to be obtained is first determined by computer computations on the basis of available analytical data. If the fertilizing value and/or the quality of the mixed product fail to comply with the requirements for its destination, a supplementary - synthetic - fertilizer dose can be added. In this way the composition of the final product can be very accurately controlled. The production process is subjected to a final check, comprising on the one hand an accurate analysis of the starting products, and a check on the composition of the final product by an external independent laboratory. The black soil is frequently used in house-building projects, road construction, landscaping and other green belt projects. In 1980 the plant, flower and tree show FLORIADE was laid out by using black soil. The organic fertilizers are mainly used as fertilizer in agriculture.

REFERENCES

(1)   INDICATIEF MEERJARENPROGRAMMA BODEM 1984-1988 (1983). Ministerie van Volkshuisvesting- en Milieuhygiene.

(2)   ZUIVERINGSSLIB IN ZUID-HOLLAND D.H.V. (1984). Raadgevend Ingenieursbureau BV. (April).

(3)   COMPOST EN ZWARTE GROND UIT ZUIVERINGSSLIB.STORA (1982). (September)
(4)   COMPUTORIZING ORGANIC WASTE TREATMENT. Karol Wojciechowicz. Internal
       publication.

DISCUSSION

A.M. Bruce
Is the process commercially viable?

J.G. ten Wolde
Yes, it is. We receive money from the water authorities who pay us to accept the raw sludge and from our customers who pay for the processed "black earth". The total cost of recycling is less than the combined sources of income and, of course, the difference is our gross profit. Incidentally, the cost of recycling is higher than that for dumping but still much cheaper than for incineration.

We are looking at many new options, for example
- composting a combined pig and domestic waste,
- collecting the ammonia produced 'and using this to make an ammonia nitrate fertiliser,
- static vs turned windrows,
- forced aeration,
- preheated aeration.

A. Kouzeli Katsiri
How does the effectiveness of static windrows compare with those that are turned?

Well, of course the land area required for static windrows is much less. However, the extent of stabilisation for a given time is much lower.

A. Wellinger
Why have you chosen to compost with wood chips and are they of special size?

We have used wood chips to open up the pile to allow for better air recirculation. The chips are not really of a special size but just those that we can obtain conveniently. The chips are waste products from the wooden shoe and paper industries in Belgium and Germany.

L.P. Savelkoul
You say you are a commercial company. Do you know how large the market for your products is likely to be? What future trends do you see, for example, inclusion of animal wastes in liquid sludges?

The market is growing all the time and has increased from 15 000 to 60 000 tons/annum over the last 3 years. Certainly, we are looking at combining animal wastes with the domestic sludge. Liquid sludges are difficult to handle though as there are high transport costs and problems with storage.

A.M. Bruce
Have you considered selling "black earth" products direct to householders?

That is one option that we may take.

SLUDGE MANAGEMENT BY THERMAL CONVERSION TO FUELS

H.W. CAMPBELL and T.R. BRIDLE
Environment Canada
Environmental Protection Service
Wastewater Technology Centre

Summary

The philosophy of sludge management in the sewage treatment industry must respond to changes in processing costs and environmental requirements. As overall costs increase, the efficiency of sludge management must be increased by either upgrading the existing scheme or introducing new technology. Both of these approaches are discussed briefly in the paper and it is pointed out that low temperature conversion of sludge to fuel, appears to have considerable potential as a viable new technology.

Experimental work carried out at Environment Canada's Wastewater Technology Centre used batch and continuous reactors to evaluate this technology at bench-scale. Tests on a number of mixed raw sludges (primary + waste activated) resulted in the following yields: oil, 22 to 25%, char, 50 to 60%, non-condensible gas, 10 to 12% and reaction water 5 to 12%.

The impact of a number of sludge treatment alternatives, including the conversion of sludge to oil, are discussed with respect to energy efficiency, flexibility, and public acceptance. The future plans of Environment Canada for the development and demonstration of sludge to oil technology are also discussed.

1. INTRODUCTION

Sewage sludge is an unavoidable by-product of wastewater treatment and roughly one tonne of sludge is generated per 4 540 m^3 (1.0 million gallons) of wastewater treated (Black and Schmidtke, 1979). Disposal of sewage sludge is an expensive process and normally constitutes up to 50% of the total annual costs for wastewater treatment. It is estimated that over 500 000 tonnes of dry sewage sludge are produced annually in Canada and disposed of at a total cost of about $104 million (Bridle, 1982). Major sludge disposal options used in Canada include agricultural utilization, landfilling and incineration, with estimated disposal costs ranging from about $126/tonne for agricultural utilization to over $300/tonne for incineration (Simcoe Engineering, 1980).

The problems of sludge disposal are expected to intensify in the future due to a number of factors. The total cost of sludge disposal will increase as the quantity of sludge to be disposed of, increases. The fraction of sludge disposed of by agricultural utilization will probably decrease due to the problem of finding suitable land within a reasonable distance of large population centres. Potential restrictions on the loading rates for sludges with high levels of heavy metals may also contribute to a decline in this practice. The feasibility of landfilling will decrease as public opposition to the licensing of new disposal sites continues to grow.

It would appear that the immediate future in sludge disposal technology will feature a trend towards more complex systems, which also implies higher costs. This will be especially true for the large urban areas where the impact of the above factors will be most keenly felt. In order to control costs, alternative resource recovery oriented solutions based on either the upgrading of current technology or the implementation of new technology, are needed.

2. TECHNOLOGY REVIEW

Upgrading current technology has been primarily concerned with improvements in dewatering and incineration. The trend has been towards considering sludge treatment from an overall systems design approach. The incentive for upgrading has been provided by the large increases in energy costs over the last ten years, and the realization of how energy inefficient most incinerator installations were. In 1973 when natural gas cost 3.6 cents/m^3, it was economically justifiable to incinerate sludge cakes of only 15 to 18% total solids. Today, natural gas costs are in the order of 18 cents/m^3, and more cost-effective methods of accomplishing moisture removal must be implemented.

Modifications which have been designed either to increase cake solids or improve the amount of energy recovered from an incinerator system include the following:
1. Upgrading from vacuum filters to belt or membrane presses.
2. The use of incinerator off-gases to dry a portion of the sludge stream.
3. Modification of incinerator burners to use digester gas as fuel.
4. Modifications to multiple-hearth incinerators to allow very dry cakes to be introduced on an intermediate hearth.
5. Waste heat recovery to generate steam, which can in turn be used to drive other processes such as thermal conditioning.

In Ontario, the improvements described above are either still under construction or have only been in operation for a relatively short period of time. Consequently, reliable information is not yet available with respect to the cost of the modifications or their effect on the energy balance of the system.

Alternative technologies also offer the potential for improved energy recovery and decreased cost. One of the most advanced is the Hyperion Energy Recovery System currently being installed in Los Angeles (Haug and Sizemore, 1983). This system comprises digestion, dewatering, Carver-Greenfield dehydration and starved air fluid bed incineration of the sludge derived fuel. It is estimated that processing 366 tonnes of raw sludge per day will generate a total of 25 MW of electricity with 10 MW available for sale back to the local utility. The complete plant is expected to be in operation by late 1985.

Sludge liquefaction has been reported as a viable method for energy production from sewage sludge, but the technology is still in its infancy. Researchers from Battelle Northwest Laboratories have reported on a sophisticated process consisting of sludge alkaline pretreatment and subsequent autoclaving at 320°C for one hour at $\approx$ 10 000 kPa under an Argon atmosphere (Molton, 1983). This process produces oil, asphalt and char, with oil yields ranging up to 15% (on a total sludge solids basis). The technology is currently being evaluated at pilot scale.

Another approach to liquefaction has been reported by Kranich et al (1980). They processed both raw and digested dry sludge with a carrier oil in an autoclave at temperatures ranging from 396-420°C under hydrogen at 10 000-13 000 kPa. Oils and asphaltenes were produced, with oil yields of up to 30%.

The basic concept of low temperature pyrolysis of sewage sludge to produce fuel products has been known for many years (Shibata, 1939). Recently, German researchers have made significant advances in understanding the mechanisms by which sludge is converted to oil (Bayer and

Kutubuddin, 1982). They heated dried sludge to 300-350°C in an oxygen free environment for about 30 minutes. The researchers postulated that catalysed vapour phase reactions converted the organics to straight chain hydrocarbons, much like those present in crude oil. Analysis of the product confirmed that aliphatic hydrocarbons were produced, in contrast to all other processes which produce aromatic and cyclic compounds, whether utilizing sludge, cellulose or refuse as the substrate. The German researchers have demonstrated oil yields ranging from 18-27% and char yields from 50-60%. The oil had a heating value of approximately 39 MJ/kg and the char about 15 MJ/kg.

Evaluation of the above technologies suggested that low temperature conversion of sewage sludge to liquid and solid fuels held the most promise of being viable at full scale. Environment Canada contracted with Dearborn Environmental Consultants to evaluate this technology, at bench scale, in both batch and continuous modes. The objective of this phase of the study was to generate process design information, for the development of cost data to assess the potential of the technology at full scale. The final phase of the program will be demonstration of the technology at full scale.

3. BENCH-SCALE STUDIES

3.1 Methodology

Raw sludges containing a mixture of primary and waste activated sludge (WAS) from three different sewage treatment plants in Ontario were evaluated during this study. All sludges were oven dried (70°C) to about 90-95% solids. Analyses of these sludges indicated they were similar in composition with volatile solids of 55-64%, calorific values of 15-18 MJ/kg and carbon contents of 24-36%.

The bench-scale studies were carried out using a batch reactor and a continuous flow system. The batch reactor, a pyrex tube 70 mm in diameter and 720 mm in length, was heated in a three-zone Lindberg furnace under a nitrogen atmosphere (Figure 1). Off-gases were condensed in a trapping system using ice as the coolant. Non-condensable gases (NCG) were vented from the system. A run was conducted by charging 550 g of dried sludge into the reactor and deaerating with nitrogen. Once operating temperature had been reached, the nitrogen purge rate was reduced. When all visible signs of reaction (i.e., gas/oil flow) ceased, the heat was switched off and the nitrogen purge rate was increased for approximately 30 minutes. The system was dismantled and the char, oil and reaction water collected and stored for analysis. Oil/water separation was achieved using a separatory funnel.

A photograph of the continuous reactor system is shown in Figure 2. It comprises a stainless steel shell, 50 mm internal diameter by 1 000 mm long, fitted with a sludge/char conveying system and is heated using the same furnace as for the batch reactor system. The reactor is subdivided by a helical gas seal, into a volatilization zone and a char/gas contact zone. The reactor system is designed for dry sludge feed rates of up to 1 kg/h. Solids retention time in the reactor is controlled by varying both the sludge feed rate and reactor inventory. Sludge is fed to the reactor by a calibrated screw conveyor and travels through the reactor by means of the reactor conveyor. Volatilized material is withdrawn in the first zone and can be contacted with the char in both co-current and counter-current modes in the second stage. Product vapours are condensed externally as per the batch reactor system. Inert gas is used to purge the system of oxygen and the operating pressure is generally less than 2 000 pascals. Prior to collection of experimental data the system was allowed to reach thermal and chemical equilibrium by operating for at least three solid retention times (SRT).

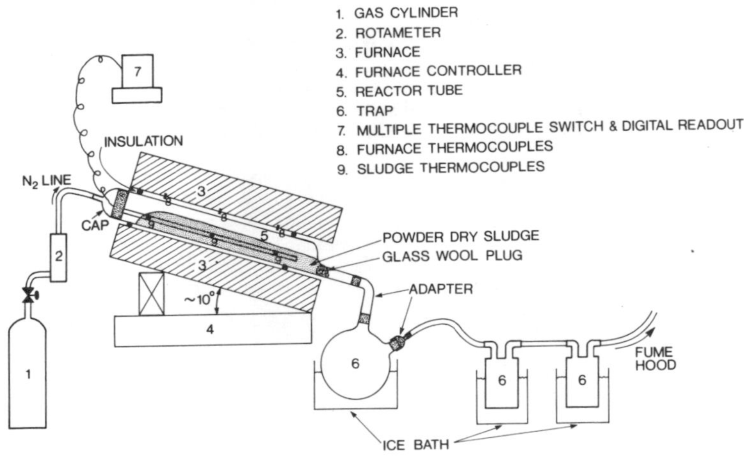

Figure 1. Batch experimental equipment

Figure 2. Continuous reactor system

3.2 Results

The results achieved with both the batch and continuous reactor systems are presented in detail in other publications (Bridle and Campbell, 1983, 1984). A summary of typical results are presented in Table 1. All the data are expressed on a dry solids basis (corrected for the normal 4-7% moisture present in the sludge) and the calorific values are expressed on a total solids basis (not corrected for volatiles).

Table 1. Typical Operating Conditions and Results

	Batch	Continuous
Feed Rate (g/h)	NA	750
Temperature (°C)	450	450
OIL		
Yield (%)	22.3	25.4
Viscosity (cstks)	>214	33.7
Calorific Value (MJ/kg)	38.9	36.7
CHAR		
Yield (%)	54.6	61.1
Calorific Value (MJ/kg)	9.4	6.2
NCG		
Yield (%)	12.1	11.1
Calorific Value (MJ/kg)	NM	5.8
REACTION WATER		
Yield (%)	11.0	5.0

NA = Not Applicable
NM = Not Measured

In general, the results from the batch and continuous systems are quite comparable. The oil and char yields are slightly higher from the continuous unit but it is difficult to say whether the difference is statistically significant. The lower calorific value of the char from the continuous reactor is the result of the lower carbon content. The most obvious difference between the products from the two systems is the oil viscosity. The oil from the batch runs was solid at room temperature (>214 centistokes), while the oil from the continuous runs was fluid at room temperature (33.7 centistokes). The batch system generated approximately twice the amount of reaction water as did the continuous one.

Although most observations are valid for either batch or continuous results, the following discussion will be limited to only the continuous system due to the fact that the experimental data base is much larger.

The yield of individual products and the split between products is a function of operating temperature. At low temperatures, the product split tends towards the formation of char. As the temperature increases to the optimum, the oil yield increases (8 to 25%) while the char yield decreases (75 to 61%). Above the optimum temperature the oil yield begins to decrease as conditions favour the formation of increasing quantities of non-condensable gas. Oil yield appears to be related to solids residence time but since it is impossible to separate completely the effects of SRT and char inventory, the trends are not clear. The yield of reaction water does not appear to be directly related to the operating parameters of the system. The yields of both oil and char are very much a function of the specific sludge used as the feed material.

The calorific value of the oil appears to be related to both SRT and temperature but their effects are relatively small. A range of SRTs from 10 to 24 minutes resulted in calorific values from 36 to 37 MJ/kg, while temperature variations from 300 to 500°C resulted in a range of only 36 to 39 MJ/kg. The calorific value of the char tends to decrease with increasing temperature but is relatively insensitive to changes in SRT. The magnitude of change in the energy of the char (i.e., from 10.4 to 5.5 MJ/kg), is substantially greater than that of the oil. The calorific value of the non-condensable gas increases in direct proportion to the temperature. This is due to the increase in the percentages of hydrogen, methane and ethane produced at higher temperatures.

The viscosity of the oil is a function of both SRT and temperature. Viscosity decreases as either SRT or temperature increases. The viscosity of the oil produced under a given set of operating conditions will be dependent on the source sludge used. For example, under optimum operating conditions, two different sludges produced oils with viscosities of 33.7 and 52.3 centistokes.

The elemental characteristics (C,H,N,O,S) of the oil appear to be quite stable with respect to processing conditions. Some fluctuations in the carbon content are evident, but the magnitude of change is relatively small. In general, 40 to 50% of the carbon from the sludge is recovered in the oil. Oxygen is difficult to measure because an unrealistically high oxygen result will be obtained if water is present in the oil sample. Although oxygen is generally in the range of 6-9%, levels as low as 2.7% have been obtained. The elemental characteristics of the char tend to be affected more by processing conditions than do those of the oil. As the temperature increases, the carbon, hydrogen and nitrogen in the char tend to decrease.

The pyrolytic water has a total Kjeldahl nitrogen content (TKN) of 4-6% and a total organic carbon content (TOC) of 10-18%. Based on the data currently available, there does not appear to be any relationship between TKN/TOC and operating parameters.

In general, the process has proven to be very stable, at least at bench scale. As the process variables change, the split between products also changes gradually. Similarly, the quality of the products is affected by process modifications but the magnitude of the change is small. The result of this is that over the practical range of operating conditions, the performance of the system can be described as a relatively flat plateau as opposed to a peak. In practical terms this means that if the temperature unexpectedly changes by 50°C in full-scale operation, one can expect to see this reflected in relative yields, but the process will not fail. It is also unlikely that the product qualities will change sufficiently in short periods of time to affect end uses or specifications.

4. IMPACT OF SLUDGE TREATMENT PRACTICES

The selection of the most appropriate sludge management scheme must consider many factors including capital cost, operation and maintenance cost, cost stability in the future, energy recovery/usage, environmental acceptability and the difference between new construction versus upgrading existing facilities. Other less tangibles include the desirability of capital versus O and M expenditures, the public perception of acceptable technology and the flexibility to respond to changing needs in the future. The impact of sludge handling and disposal practices can be illustrated by examining the four sludge management alternatives shown in Figure 3.

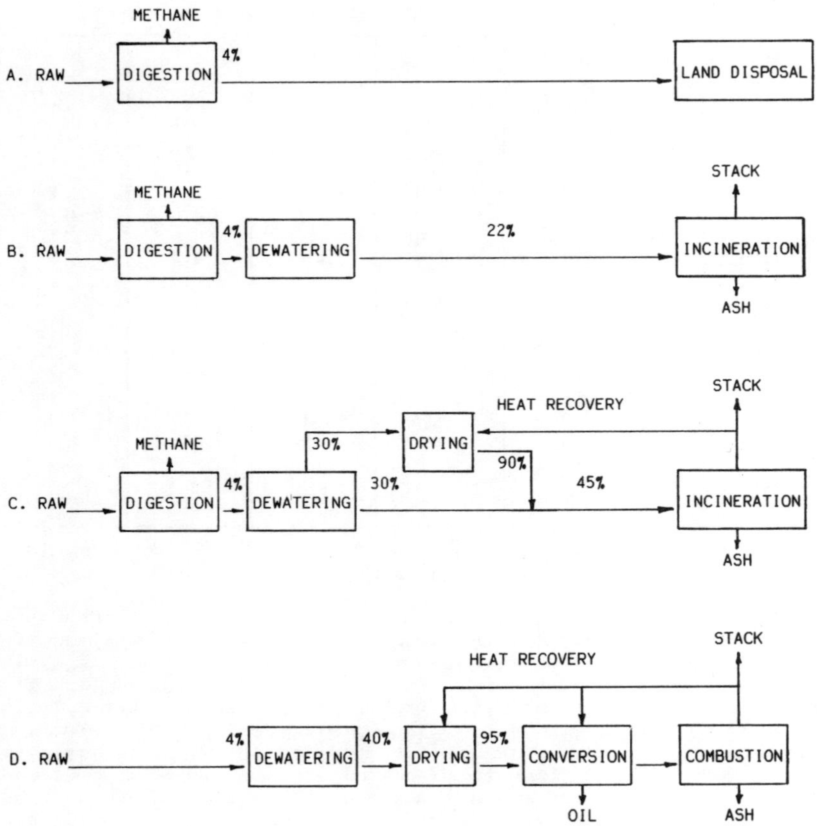

NOTE: (%) numbers refer to sludge solids concentrations at various points in the process.

Figure 3. Sludge disposal alternatives.

They range from a relatively low level of technology with many years of documented experience, to a very sophisticated system which, to date, has only been proven at bench-scale. In Table 2 the four alternatives are compared on the basis of their energy recovery potential. The analysis uses a raw sludge input, for all alternatives, of 25 dry tonnes of sludge solids per day. It is assumed that sludge, char, NCG and reaction water are combusted at 75% efficiency. In order to be consistent, the methane from digestion is only allocated 75% of its gross energy, since it still has to be burned in a furnace of some nature before the energy becomes usable.

Table 2. Energy Balance for Sludge Disposal Alternatives

ALTERNATIVE (See Figure 3)	A	B	C	D
Total Energy Available in Raw Sludge (MJ/h)	17 072	17 072	17 072	17 072
ENERGY OF PRODUCTS				
Methane (MJ/h)	4 230	4 230	4 230	0
Oil (MJ/h)	0	0	0	8 327
Char+NCG+H_2O (MJ/h)	0	0	0	5 111
Digested Sludge (MJ/h)	0	5 414	5 414	0
ENERGY REQUIREMENTS Reactor/Dryer Incinerator/Combustor (MJ/h)	0	12 797	7 580	1 739
NET ENERGY				
(MJ/h)	4 230	-3 063	2 064	11 699
(MJ/kg raw sludge)	4.06	-2.94	1.98	11.23
ENERGY RECOVERY				
(%)	24.8	0	12.1	68.5

NOTE * Net Energy = Total Energy of Products - Total Energy Requirements
    * Energy Recovery = (Net Energy/Raw Sludge Energy)x(100)

Alternative A, the application of digested sludge on agricultural land, has traditionally been the preferred method of sludge management for many municipalities. For smaller communities which have a sludge with acceptable levels of heavy metals, this is still the most cost-effective method of disposal. As communities become larger, drawbacks to this system become evident. It is difficult for large municipalities to find sufficient land within a reasonable distance of the plant on which to apply the sludge. Even if the required land is available, the number of trucks needed to transport liquid sludge may create a local traffic problem which can adversely affect public relations. As municipalities become more industrialized, the potential that the sludge may not satisfy heavy metal guidelines for land application also increases. In terms of energy recovery the system is quite efficient showing a net energy of 4.06 MJ/kg of raw sludge. The reason for this level of efficiency is that there are no large energy requirements for this alternative, although the energy required to transport the sludge to the farmland has not been accounted for. One of the major problems with methane as an energy source is that because it is difficult to store and transport, it must be utilized on-site continuously. Many sewage treatment plants find that this is not possible in summer, and consequently, flare a significant fraction of their gas. Thus, while a digester may be generating potential energy at the rate of 4 230 MJ/h, the rate of utilization over the entire year may be considerably less.

Alternative B has been a common practice for many installations. The sludge is digested, dewatered to 20 to 22% and incinerated in a multiple-hearth or fluidized bed incinerator. This alternative is significantly more expensive than Alternative A and is much less energy efficient. Reference to Table 2 shows that there is a net energy deficit of 2.94 MJ/kg of raw sludge, assuming that the incinerator has been modified to burn methane as auxiliary fuel. This inefficiency is primarily due to the energy requirements to evaporate the water (78%) in the incinerator. If the energy deficit were satisfied by natural gas, this would translate to an auxiliary fuel cost of approximately $150 000 per year. The positive features are that there is a minimum quantity of material (ash) for ultimate disposal, and the majority of heavy metals will be immobilized in the ash. Due to the high energy requirements, this system is generally being phased out.

Alternative C is one approach to maintaining the same basic system as outlined above, but upgrading it in order to improve energy efficiency and subsequently reduce operation and maintenance costs. The sludge dewatering equipment is upgraded in order to achieve 27 to 30% solids in the cake on a consistent basis. A portion of the cake is then dried to 85 to 90% solids using waste heat from the incinerator. This portion is backmixed with the rest of the dewatered cake (30% solids) such that the combined sludge feed to the incinerator is in the order to 40 to 50% solids. Depending on the volatile fraction of the sludge, the cake may or may not be autogenous. The capital cost will be higher than for Alternative B due to the addition of the dryer but the energy balance is much more positive. In this case, a net energy surplus of 1.98 MJ/kg of raw sludge is produced but it should be stressed that any net energy not utilized in the process is only available as either methane gas or hot flue gas. Both of these represent a difficult problem in terms of either storage or transportation. Another problem with either Alternative B or C is that, in many cases, the public views sludge incineration as an undesirable practice.

Either system could also be used without digestion. This would probably increase the level of cake solids achievable in dewatering and would significantly increase the available energy in the sludge going to the incineration. The potential of using methane as auxiliary fuel would be lost and the elimination of digestion would mean that a backup method for sludge disposal during incinerator down-time would also be lost. When all factors are considered, it is not clear whether it is more economical to incinerate raw or digested sludge.

Alternative D has currently only been evaluated at bench-scale, but appears to have a number of novel aspects which could make it a publicly acceptable and economical method of sludge disposal. As mentioned above, the fact that raw sludge is being dewatered indicates that cake solids in the order of 40% can be achieved using technology such as a diaphragm filter press. The energy balance in Table 2 shows the impact on recovered energy. The net energy (11.23 MJ/kg of raw sludge) is 2.75 times higher than Alternative A, and almost six times higher than Alternative C. All process energy requirements for Alternative D could be supplied by 34% of the energy in the char, NCG and reaction water. This would leave 3 372 MJ/h of energy available for other uses in the form of recovered heat, and 100% of the oil (8 327 MJ/h) would be available for sale. The primary reason for the differences in the energy balances between Alternatives C and D is in the manner of energy conversion. In C the conversion is either by biological means (i.e., digestion), or by heat recovery. Neither of these methods are as efficient as the process in D which consists of catalysed vapour phase reactions converting the lipids and proteins in the sludge to straight chain hydrocarbons.

The fact that a large percentage of the energy in Alternative D is available as oil has a significant impact on the entire sludge management philosophy. Instead of having to use recovered energy in-house, which is usually the case with methane, steam, hot flue gas, and even electricity, the energy is now in a storable, transportable and potentially saleable form. Currently, the least valuable end-use for the oil appears to be as a substitute for No.6 fuel oil. If the oil can be sold for $30/barrel, then the net revenue from a 25 tonne of dry sludge per day plant would be approximately \$365 000 per year. However, the potential exists to increase end-use value, and upgrading to a transportation fuel is a possibility.

The use of a char combustor as opposed to a sludge incinerator may also have a distinct advantage with respect to public acceptance. The combustion of char may be seen to be more analagous to burning coal, rather than incinerating sewage sludge with all of the perceived associated environmental concerns. The char combustor will be similar to an incinerator in that they will produce the same amount of material for ultimate disposal and the heavy metals will, to a large extent, be immobilized in the resulting ash. A char combustor should also be considerably smaller than an incinerator to burn sludge from an equivalent sized plant because the capacity to evaporate the large amounts of water which are normally present in sludge, is not required.

Sludge management by conversion to oil will be most advantageous when considering the construction of new plants. Since the major source of the raw material for oil is the biomass from the biological treatment plant, the ideal feedstock for the process is raw sludge. This eliminates the need for the construction of digesters and can be translated into savings in capital investment for new plant construction. The process is also compatible with high-rate biological processes. One of the drawbacks of high-rate processes has always been the additional quantities of sludge generated and the corresponding increases in sludge disposal costs. By converting the excess sludge to oil, any additional cost resulting from the increased quantity requiring treatment, should be offset by the increased volume of oil produced.

5. FUTURE TRENDS

The approach to sludge management in the sewage treatment industry has changed over the past few years and will continue to change in the foreseeable future. Considerations of energy, cost and availability have shown that many sludge handling options, which were considered state-of-the-art only a few years ago, are no longer economically justifiable. Sludge management schemes must be designed with an overall systems approach to both the solids flow and energy efficiency. The conversion of sludge to oil offers an áttractive alternative to sludge options currently in operation. The conversion technology is estimated to be at least comparable to incineration in terms of capital cost and minimal negative impact on the environment. The most important advantage is that the recovered energy is in the form of oil which is storable, transportable and potentially saleable.

It is estimated that by 1985, 350 000 tonnes of sewage sludge will be incinerated annually in Canada. Thermal conversion of this sludge could produce 700 000 barrels of oil per year, with a market value of at least \$21 million.

Environment Canada's long term plans are to demonstrate this technology by construction of a 25 tonne per day facility. The first phase of this demonstration study is currently underway with the development of the pre-engineering data, identification of suitable sites and a thorough assessment of the process economics. This phase of the program is being conducted under contract by Zenon Environmental Inc. and Xytel Energy Ltd.

REFERENCES

(1)   BAYER, E. and KUTUBUDDIN, M. (1982). "Low Temperature Conversion of Sludge and Waste to Oil". Proceedings of the International Recycling Congress, Berlin, West Germany.

(2)   BLACK, S.A. and SCHMIDTKE, N.W. (1979). "Practises and Trends in Sewage Sludge Utilization and Disposal", presented at 1st Workshop on Canadian-German Cooperation, Burlington, Ontario.

(3)   BRIDLE, T.R. (1982). "Sludge Derived Oil: Wastewater Treatment Implications", Env. Tech. Letters Vol.3, pp.151-156.

(4)   BRIDLE, T.R. and CAMPBELL, H.W. (1983). "Liquid Fuel Production from Sewage Sludge", presented at the ENFOR Third Canadian Biomass Liquefaction Experts Meeting, Sherbrooke, Quebec.

(5)   BRIDLE, T.R. and CAMPBELL, H.W. (1984). "Conversion of Sewage Sludge to Liquid Fuel", presented at the 7th Annual AQTE Conference, Montreal, Quebec.

(6)   HAUG, R.T. and SIZEMORE, H.M. (1983). "Energy Recovery and Optimization: The Hyperion Energy Recovery System", presented at the International Conference on Thermal Conversion of Municipal Sludge, Hartford, Connecticut.

(6)   KRANICH, W.L., GURUZ, K. and WEISS, A.H. (1980). "Hydroliquefaction of Sewage Sludge". Proceedings of the National Conference on Municipal and Industrial Sludge Utilization and Disposal, Washington, D.C.

(7)   MOLTON, P.M. (1983). "Batelle-Northwest Sewage to Fuel Oil Conversion", presented at the International Conference on Thermal Conversion of Municipal Sludge, Hartford, Connecticut.

(8)   SHIBATA, S. (1939). "Procède de Fabrication d'une Huille Combustible à Partir de Boue Digérée". French Patent 838,063.

(9)   Simcoe Engineering. (1980). "Sludge Management Study for the Regional Municipality of Halton", Pickering, Ontario.

DISCUSSION

P.J.W. ten Have
Is it usual to achieve sludge with a solids content as high as 40% in practice?

H.W. Campbell
Yes, with a diaphragm press, you can achieve levels up to about 44%.

Do the heavy metals present in the sludge affect the "oil-making" process?

Yes, they can have an effect. The heavy metals present tend to remain in the char formed and are in the same form as produced by incineration. On the other hand, the organics present in the sludge tend to be concentrated in the oil. This can impose restrictions on the outlets for the oil. However, for many users, for example, firing a cement kiln, it would not cause any difficulty.

H. Scheltinga
I am surprised how simple the process is.

Yes it is very simple especially as it works at atmospheric pressure. The heat recovery system could, however, be quite complex. The process is likely to be capital intensive though the overall economics of the process could be quite good.

A.M. Bruce

You have used raw sludge in
your process up to now. Could
you use anaerobically digested
sludge which had been used to
produce biogas?

Yes, you could but as some of the
carbon content had been converted
into biogas, then the yield of oil
would only be about one half.
Incidentally, the oil produced in
this case is superior as it tends
to be much less viscous.

P. Balmer

How do the capital costs of your
process compare with incineration?

The accepted costs for incineration
are about US $300/ton. Our costs
will be competitive with this and
could be as low as one half.
Certainly the oil produced is very
valuable and is almost as good in
quality as diesel.

ACCUMULATION OF REFRACTORY 4-NONYLPHENOL DURING MESOPHILIC ANAEROBIC SLUDGE STABILIZATION

M. TSCHUI, P.H. BRUNNER and W. GIGER
Swiss Federal Institute for Water Resources and Water Pollution Control
(EAWAG), Federal Institute of Technology,
CH-8600 DÜBENDORF, Switzerland

Summary

Anaerobically stabilized sewage sludges contain high concentrations of toxic 4-nonylphenol. Continuous laboratory-scale experiments under controlled mesophilic anaerobic conditions showed that the enrichment of 4-nonylphenol is largely due to the transformation of 4-nonylphenol mono- and diethoxylate which are metabolites of the widely used nonionic surfactants of the 4-nonylphenol polyethoxylate type.

1. INTRODUCTION

4-nonylphenol polyethoxylates (Figure 1A) are widely used nonionic surfactants which occur in municipal wastewaters (Ahel and Giger, 1985b). 4-nonylphenol (NP), 4-nonylphenol mono- and diethoxylate (NP1EO, NP2EO; Figure 1) were detected in biologically treated sewage effluents (Stephanou and Giger, 1982; Ahel and Giger, 1985a). In activated sludges NP1EO and NP2EO as well as NP were found in the range of 0.09 to 0.15 g/kg dry matter of sludge. In anaerobically digested sludges 4-nonylphenol was detected in extraordinarily high concentrations of 0.45-2.5 g/kg dm (Giger et al., 1984).

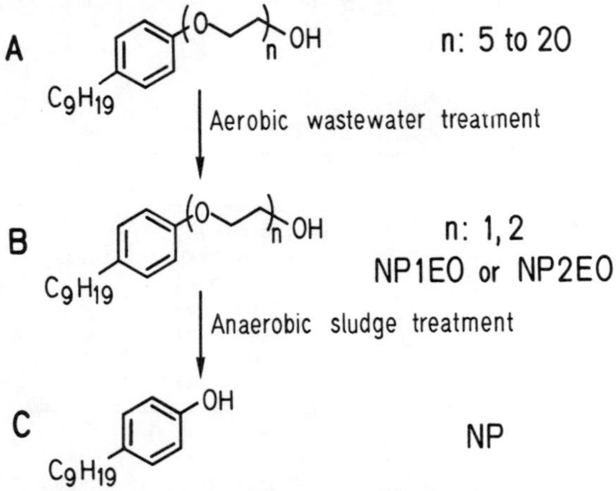

Figure 1. Degradation of 4-nonylphenol polyethoxylates (A) during wastewater and sludge treatment (Giger et al., 1984)

The following degradation pathway has been proposed (Figure 1, Giger et al., 1984): The 4-nonylphenol polyethoxylates are being degraded to di- and monoethoxylates during aerobic sewage treatment (Rudling and Solyom,

1974). The resulting ethoxylates are less biodegradable and less hydrophilic than the unaltered surfactants. These metabolites are sorbed by the lipophilic sludge-flocs. In the subsequent anaerobic sludge treatment, the mono- and diethoxylates are further degraded to the 4-nonylphenol.

We report here an investigation into the formation of 4-nonylphenol during anaerobic mesophilic sludge stabilization in order to examine the above hypothesis of the anaerobic formation of 4-nonylphenol.

2. METHODS

Two sets of experiments were performed: the anaerobic formation of 4-nonylphenol was examined in a continuous laboratory fermenter (Figure 2), Model LF 20 SK, FZ 005 (Chemap AG CH-8604 Volketswil). The substrate was fed pneumatically by a piston pump. The amount of digested sludge leaving the fermenter was controlled by a balance which kept the weight of the reactor constant.

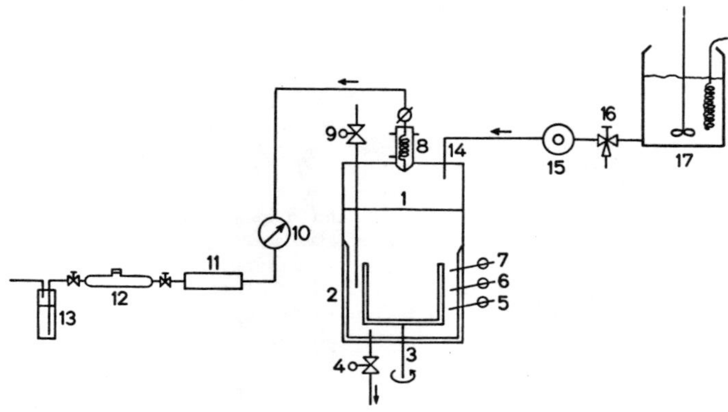

Figure 2. Apparatus for continuous anaerobic mesophic digestion.
Legend:

1	reactor	9	sample port
2	heater	10	gas meter
3	stirrer	11	water trap (silica)
4	pneumatic valve, sludge outlet	12	gas sampling
5	pH-electrode	13	water flask
6	redox electrode	14	sludge inlet
7	thermocouple	15	feed pump
8	water trap (condensation), gas outlet	16	sample port
		17	substrate, stirred and cooled

The anaerobic degradation of 4-nonylphenol was investigated by batch experiments in 3 and 5 l double wall roundflasks.

All experiments were performed at 35°C. Parameters analyzed included pH, volatile acids, alkalinity, NH_4-nitrogen, Kjeldahl-nitrogen, dry matter, loss on ignition, COD, gas volume and -composition, total carbon and nitrogen, 4-nonylphenol, 4-nonylphenol mono- and diethoxylate.

The substrate which was used for all experiments was raw, dried sewage sludge from the sewage treatment plant of Altenrhein, Switzerland. This sludge was further dried in the laboratory at 60°C for 2 days. It was thoroughly mixed (concrete mixer), pulverized (sieve <1 mm), and stored in portions at -20°C. The following parameters characterize this sludge:

dry matter   96.8 ± 0.6 g/100 g sludge
loss on ignition   56.5 ± 0.5 g/100 g dry matter
Total carbon   29.9 ± 0.8 g/100 g dm
Total nitrogen   3.2 ± 0.1 g/100 g dm

For the digestion experiments the dried sludge was suspended in water to a dry matter content of 4 g/100 g sludge.

The quantitative determination of 4-nonylphenol and the corresponding mono- and diethoxylate was performed by steam distillation/extraction with cyclohexane and normal-phase high-performance liquid chromatography (HPLC) (Ahel and Giger, 1985a). For quantification, 2, 4, 6-trimethylphenol was added as an internal standard. HPLC was carried out on a Perkin Elmer Series 4 liquid chromatograph using a bonded-phase aminosilica column (Hypersil-APS, 3 µm, 60 x 4 mm id, Knauer, Berlin). n-Hexane and n-hexane/2-propanol 85/15 were used as solvents for gradient elution. A Kratos Spectraflow UV/VIS-absorbence detector (Model 773) set at a wave length of 277 nm was used.

3. RESULTS

Our experiments confirm the formation of 4-nonylphenol by anaerobic mesophilic digestion of sewage sludge containing 4-nonylphenol mono- and diethoxylates.

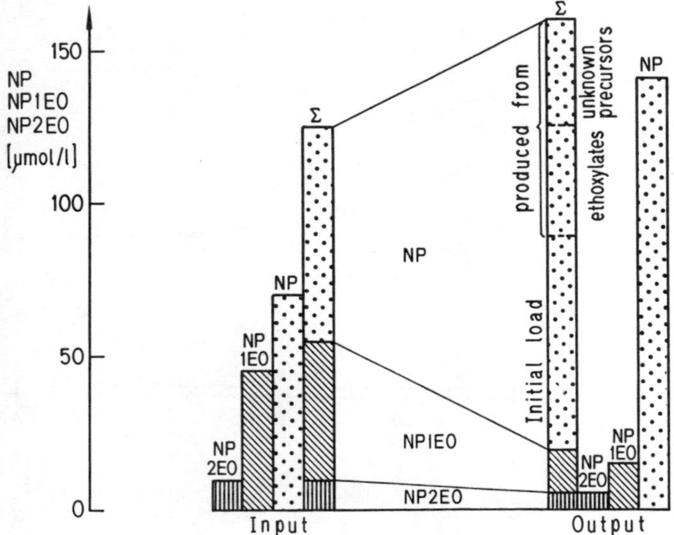

Figure 3. Mass balance for the continuous digestion of 4-nonylphenol (NP), 4-nonylphenol monoethoxylate (NP1EO) and 4-nonylphenol diethoxylate (NP2EO) at 15 days residence time.

When the fermenter was operated at a residence time of 20 days, approximately 60% of NP1EO and NP2EO were degraded and NP increased by a factor of 2 (Figure 3). Assuming that the metabolized NP1EO and NP2EO were transformed to NP, a kinetic model revealed that there must be additional other educts which contribute approximately 40-50% of the production of the NP formed. Higher polyethoxylates are probably no significant precursors since they are assumed to be mainly degraded in the aerobic treatment step.

When the fermenter was operated at 15 days residence time, degradation and accumulation were the same as for 20 days.

For further confirmation, the fermenter of steady state operation was spiked with 4 000 µmol NP1EO, 1 088 µmol NP2EO and 368 µmol NP3EO. Subsequently, the concentration profiles for the three ethoxylates and 4-nonylphenol were monitored for 60 days. Figure 4 shows the results for the measured values (dots) and the calculated values (lines). The calculated values were obtained using the initial concentrations, the washout of the products at a residence time of 15 days and the first order reaction rates obtained from the steady state experiments. Measured and calculated values agree quite well, thus confirming that the anaerobic mesophilic digestion of NP1EO and NP2EO yields 4-nonylphenol. This experiment also shows the remarkable capacity of an anaerobic fermenter to cope with single loads of NP1EO and NP2EO.

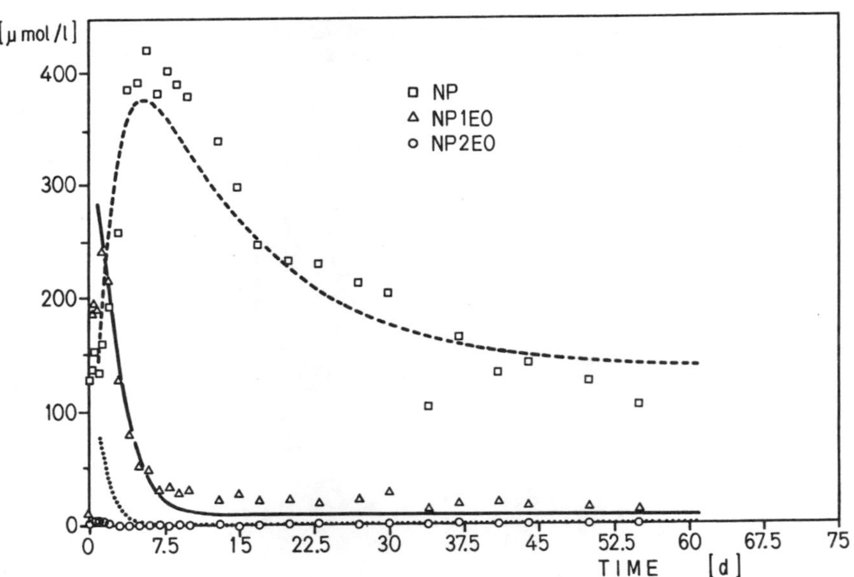

Figure 4. Concentrations of 4-nonylphenol (_ _□_ _), 4-nonylphenol mono-ethoxylate (___Δ___) and 4-nonylphenol diethoxylate (..○..) during continuous anaerobic mesophilic sludge treatment.

□ , Δ , ○   measured concentrations

_ _ _, _____, ...   calculated concentrations.

At t ≈ 0, the digester (15 l) was spiked with 4 000 µmol NP1EO, 1 088 µmol NP2EO and 368 µmol NP3EO.

It was not possible to determine if 4-nonylphenol is strictly refractory under anaerobic conditions by the batch experiments which were performed. Therefore, the models used assumed no further degradation of 4-nonylphenol in the continuous fermenter.

Since the work presented here verifies the formation of refractory materials during anaerobic mesophilic stabilization, the question arises, if other stabilization processes yield the same products. Preliminary results of a field investigation into the fate of 4-nonylphenol polyethoxylates during aerobic and anaerobic mesophilic and thermophilic stabilization indicate that the degradation products formed by these processes are significantly different (McEvoy et al., 1985).

REFERENCES

(1)   AHEL, M. and GIGER, W. (1985a). Anal. Chem. 57, in press.
(2)   AHEL, M. and GIGER, W. (1985b). Anal. Chem., submitted.
(3)   GIGER, W., BRUNNER, P.H. and SCHAFFNER, C. (1984). Science, 225, 623.
(4)   McEVOY, J., GIGER, W. and BRUNNER, P.H. (1985), in preparation.
(5)   RUDLING, L. and SOLYOM, P.A. (1974). Water Res. 8, 115.
(6)   STEPHANOU, E. and GIGER, W. (1982). Environ. Sci. Technol. 16, 800.

This work was supported by the Swiss Department of Commerce (Project COST 68). We thank H. Moench, C. Schaffner, M. Ahel for their technical assistance and G. Hamer for the use of fermenter, Ch. Schumacher for the typing and H. Bolliger for the illustrations.

DISCUSSION

A.M. Bruce
Is 4-nonylphenol very toxic?

M. Tschui
Yes, it can be very toxic, for example, to daphnia and to aquatic plants. The toxicity is largely removed during aerobic digestion.

F. Langeweg
Is it readily biodegradable under aqueous conditions?

Yes, I have been able to measure the biodegradability of 4-nonylphenol in sludges and effluents under aqueous/aerobic conditions.

P.W. ten Wolde
What measurements of toxicity have you used?

H. Scheltinga
You need good analytical methods for the estimation of 4-nonylphenol before you start measuring the toxicity. Working Parties 2 and 5 have developed suitable methodology. This is a very important first step.

CONCLUSIONS

P. BALMER
Chalmers University of Technology, GOTHENBURG - Sweden
T.C. CASEY
University College, DUBLIN - Ireland

The presentations covered a range of topics including dewatering processing, disposal and toxic detergent residues.

Dr. Colin presented a review of electrochemical enhancement of water/particle separation in aqueous solutions and of the application of such processes in the dewatering of slurries. Although his review showed evidence of scientific interest in these processes over the past decade, it is clear that their application in the field of sludge dewatering is not yet feasible.

Mr. Spinosa presented laboratory findings related to the selection of polyelectrolytes for sludge conditioning purposes using charge density and molecular weight as classifying polymer characteristics. His results, together with the findings reported by other workers in this field, should help to provide a basis for a more rational approach to screening of candidate polymers.

Dr. Loll reviewed the general performance characteristics of presently available dewatering equipment and set out a target specification against which a new continuous high pressure belt press had been developed. He presented performance data from a prototype machine which achieved a dewatered cake solid in excess of 40%. The capability of producing high cake solids in a continuous dewatering process is a significant technological advance. The performance of this new machine in extended trials will be watched with considerable interest by those engaged in the sludge dewatering business.

Mr. Paulsrud presented details of Norwegian experience of the aerobic thermophilic digestion process, gained from a pilot plant study using a Danish proprietary system. His reported performance characteristics were in general agreement with operating data from other countries. In particular, he reported that effective hygienization could be achieved at a retention time of about 3 days. However, incomplete hygienization was observed with short retention times even though sludge temperatures above 60°C had been achieved.

Mr. Tschui presented experimental evidence of the accumulation of the toxic detergent derivative, 4-nonylphenol in mesophilic anaerobic digestion processes. It is clear from his findings that more information is required on the production and fate of such refractory compounds in sludge treatment processes.

Mr. Campbell reported on continuing Canadian studies on oil extraction from sludge by pyrolysis. The technical feasibility of this process has been demonstrated at laboratory level and desk studies show the process to be economically competitive with incineration with more favourable energy yields. The process now awaits scale-up evaluation.

Mr. ten Wolde described a large-scale composting operation in the Netherlands which combines digested sludge with sand to form 'black soil' and with other waste residues to form an organic fertiliser. The operation is characterised by a market-oriented approach which aims to produce sludge-containing products which meet the requirements of the user.

Mr. ten Have reported on the technical measures which are being evaluated for the processing of animal manures in the Netherlands, where animal manure production exceeds the amount which can be recycled in agriculture. It is intended to introduce legislation in 1986 to control the rate of fertiliser application to agricultural land. In outlining the various treatment options Mr. ten Wolde drew attention to the difficulty of disposing of the ultimate residues and to the economic implications of high treatment costs.

LIST OF PARTICIPANTS

BALMER, P.
Chalmers University of Technology
Water Supply and Sewerage
Engineering
Fack
S - 41296 GÖTEBORG

BEKER, D.
RIVM
Antonie van Leeuwenhoeklaan 9
NL - 3720 BA BILTHOVEN

BRINKHORST, M.A.
BKH
Postbox 93224
NL - 2509 AE s'GRAVENHAGE

BRUCE, A.M.
Water Research Centre
Elder Way
UK - STEVENAGE, HERTS SG1 1TH

CAMPBELL, H.W.
Waste Water Technology Centre
Environmental Protection
Service Environment Canada
P.O.Box 5050
CANADA - BURLINGTON,
            Ontario L7R 4AE

CASEY, T.J.
University College Dublin
Department of Civil Engineering
Earlsfort Terrace
IRL - DUBLIN 2

COLIN, F.
Institut de Recherches
Hydrologiques de Nancy
10, rue Ernest Bichat
F - 54000 NANCY

DORUSSEN, H.L.
BKH
Postbox 93224
NL - 2509 AE s'GRAVENHAGE

DUVOORT VAN ENGERS, L.E.
RIVM
Antonie van Leeuwenhoeklaan 9
NL - 3720 BA BILTHOVEN

KAMPF, R.
TNO
Afd Water en Bodem
NL - 2628 VK DELFT

KARPER, R.
Rijksinstituut voor de Zuivering
van Afvalwater
Maerland 5/6
Postbus 17
NL - 8200 AA LELYSTAD

KOUZELI-KATSIRI, A.
National Technical University
Chair of Sanitary Engineering
5 Iroon Polytechniou
St Zografou
GR - ATHENS 624

LANGEWEG, F.
RIVM
Antonie van Leeuwenhoeklaan 9
NL - 3720 BA BILTHOVEN

L'HERMITE, P.
DG XII/G1
Environmental Protection and
Climatology
200, rue de la Loi
B - 1049 BRUSSELS

LOLL, U.
Abwasser-Abfall-Aquatechnik
Heinrichstr. 10
D - 6100 DARMSTADT

LOTITO, V.
IRSA
Reparto Sperimentale di Bari
Via Francesco de Blasio 5
I - 70123 BARI

NEWMAN, P.J.
Water Research Centre
Medmenham
UK - MARLOW, BUCKS SL7 2HD

NOUWEN, P.
Ministerie van VROM
Postbox 450
NL - 2260 MB LEIDSCHENDAM

OCKIER, P.
Zandvoordestraat 375
B - 8400 OOSTENDE

PAULSRUD, B.
Aquateam
Norwegian Water Technology
Centre A/S
NORWAY

PUOLANNE, Y.J.
National Board of Waters
Pohjoinen Rautatiekatu 21B
P.O. Box 250
SF - 00101 HELSINKI 10

SANTORI, M.
IRSA
Via Reno 1
I - 00198 ROME, CAP

SAVELKOUL, L.P.
Public Water Laboratory
Municipality of Amsterdam
Wibautstraat 3
NL - 1091 GH AMSTERDAM

SCHELTINGA, H.M.J.
Staatstoezicht op de
Volksgezondheid
Hoofdinspectie voor de Hygiene van
het Milieu
Pels Rijckenstraat 1
NL - 6800 DR ARNHEM

SCHLOSSER, P.J.J.
Zuiveringsschap
Postbox 314
NL - 6040 AH ROERMOND

SPINOSA, L.
IRSA
Reparto Sperimentale di Bari
Via Francesco de Blasio 5
I - 70123 BARI

TEERINK, J.
Hoogheemraadschap van Rijnland
Postbox 156
NL - 2300 AD LEIDEN

TEN HAVE, P.J.W.
Government Agricultural Waste Water
Service (RAAD)
Kemperbergerweg 87
NL - 6816 RM ARNHEM

TEN WOLDE, P.J.
Rutte Recycling BV
Osdorperweg 578
NL - 1067 SZ AMSTERDAM

TSCHUI, M.
EAWAG
CH - 8600 DÜBENDORF

VAN DEN BERG, J.
TAUW
Postbox 479
NL - 7400 AL DEVENTER

VAN HELVOORT, P.C.A.M.
Hoogheemraadschap van de
uitwaterende sluizen in
Kennemerland en West-Friesland
P.O.Box 15
NL - 1135 ZH EDAM

VAN OERS, R.E.M.
West Brabant River Board
Postbox 2212
NL - 4800 CE BREDA

VERHAAGEN, J.
Zuiveringsschap West Overijssel
Postbox 50
NL - 8000 AB ZWOLLE

WELLINGER, A.
INFOSOLAR c/o FAT
SWITZERLAND

WERMEUS-BUNING, W.G.
Hollandse Eilandewen Waarden
Postbox 469
NL - 3300 AL DORDRECHT

WITTEBROOD, R.J.
Prov. Waterstraat
Friesland
Postbox 1186
Nl - 8900 CD LEEUWARDEN

LIST OF AUTHORS